Sabertooth

Life of the Past   James O. Farlow, editor

SABER

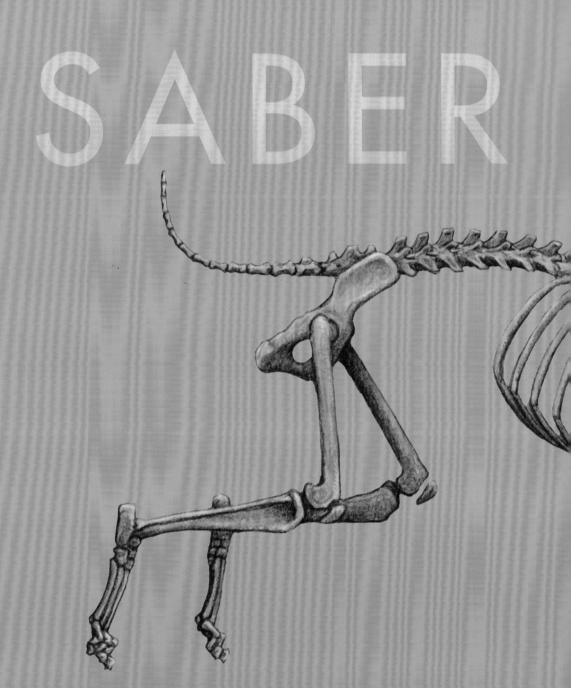

Indiana University Press   Bloomington and Indianapolis

TOOTH

Mauricio Antón

This book is a publication of

Indiana University Press
Office of Scholarly Publishing
Herman B Wells Library 350
1320 East 10th Street
Bloomington, Indiana 47405 USA

iupress.org

This book is printed on acid-free paper.

Manufactured in China

Library of Congress Cataloging-in-Publication Data

Antón, Mauricio.
Sabertooth / Mauricio Antón.
pages cm. – (Life of the past)
Includes bibliographical references and index.
ISBN 978-0-253-01042-1 (cl : alk. paper)–ISBN 978-0-253-01049-0 (ebook)
1. Saber-toothed tigers. I. Title.
QE882.C15A55 2013
569′.7–dc23
2013001962

3 4 5 6 7 27 26 25 24 23 22

FOR MY FAMILY AND IN MEMORY OF ALAN TURNER.

AND WHAT SHOULDER AND WHAT ART   COULD TWIST THE SINEWS OF THY HEART?

AND WHEN THY HEART BEGAN TO BEAT,   WHAT DREAD HAND AND WHAT DREAD FEET?

WILLIAM BLAKE, "THE TYGER"

Contents C

Foreword

FOR THE PAST THIRTY-FIVE YEARS I HAVE HAD THE PRIVILEGE OF spending time in the company of Africa's charismatic big cats—the lions, leopards, and cheetahs that my wife, Angie, and I have come to know as individuals, recording their lives in words, drawings, and photographs in the Masai Mara in Kenya, the northern extension of Tanzania's great Serengeti National Park.

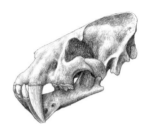

The Mara-Serengeti is an ancient land: there are rocks at the heart of the Serengeti that are more than three billion years old. Standing on a hilltop overlooking the vastness of the Serengeti's short-grass plains during the rainy season, you can witness a Pleistocene vision, the land awash with animals. Hundreds of thousands of wildebeest and zebras, tens of thousands of gazelles, and hundreds of elands and ostriches share the mineral rich grasslands. Dotted among them you can pick out the sloping backs and powerful forequarters of spotted hyenas as they amble along, the herds parting and closing again as the predators pass through or begin to hunt. Prides of lions rest up in the shade of granite outcrops known as kopjes that emerge like castles from a sea of grass. Somewhere a leopard lies recumbent along the wide limb of a giant fig tree, while a cheetah perches sphinx-like on a termite mound, looking for a gazelle fawn to chase down. This is the last place on earth where you can see scenes of such abundance, yet it's only a fragment of our planet's past animal glories.

The fossil record allows us a glimpse of other times and other creatures equally as fascinating and awe inspiring as anything seen today, times when there were many more members of the cat family searching for prey among wild landscapes across the globe.

We are mesmerized by predators. There is a mixture of awe and fear, a reminder at some primal level of the time 2 million years ago when our ancestors emerged from the forest edges into the sunlight of the African savannahs, scavenging and killing prey for themselves. To do this they had to find ways of competing with the great cats and hyenas of that epoch. Little wonder, then, that we fear the large predators for their power while admiring them for their strength and courage. This ambiguous relationship between hunter and hunted is echoed in the hauntingly beautiful cave art of Lascaux and Chauvet in Europe—an artistic tradition that Mauricio Antón so admirably continues with the artwork in his fascinating and informative *Sabertooth*. It takes a skilled observer with imagination to bring the past to life.

Who hasn't felt a thrill and fascination at the mention of sabertooths? Cartographers of ancient times inscribed "here be dragons" on early maps, conjuring up vivid images of giant reptiles living deep in the heart of Africa, in the same way that sabertooth tigers (as they were sometimes referred to) were the highlight of the Boys Own magazines and comics of my childhood. Something for youngsters to fantasize about; ripping adventures played out in the wilds of Africa and beyond.

I first came across Mauricio Antón's eye-catching draftsmanship in a copy of his *The Big Cats and Their Fossil Relatives* when Angie and I were researching a series of books on Africa's big cats to accompany the popular television series *Big Cat Diary*. Mauricio's beautifully crafted drawings and paintings allowed us to step back in time to a very different era. If a love of the African savannahs has driven our own passion for wilderness and adventure, then imagine the thrill of taking a safari through the Pleistocene landscape of a million years ago–or further back still, to the Miocene of 20 million years ago that heralded the advent of the extinct relatives of our modern big cats.

The world of fossils and prehistoric life must by its very nature remain part of our imagination–something ancient and to a degree unfathomable. It takes the vision of an artist and the dogged determination of a detective, combined with the highest understanding of our current knowledge of anatomy and animal behavior, to conjure up illustrations that are both believable and awe inspiring. This is Mauricio's gift, and I particularly love his panoramas: colorful renditions of complete landscapes that suggest a dynamic and living storyboard of creatures and events from tens of thousands of years ago–millions, in some cases.

The largest and most famous of the sabertooth cats is *Smilodon*, an animal that was larger and more powerful than the largest living tiger and that roamed the American landscapes as recently as ten thousand years ago. Its saber teeth–curved, dagger-like canines–have been the source of speculation and inquiry into why such fearsome yet fragile weapons evolved and how they were used, questions that Mauricio Antón attempts to answer in this book.

The natural world we live in is constantly evolving, molded by climate, soil, and competition between species through the process of natural selection. The wonder of the wild animals, plants, and trees we see today are its creation, as are the remnants of times past in the form of fossils. From these ancient fragments and a thorough knowledge of all these processes combined with the findings of the very latest DNA technology, Mauricio is able to take us on a journey of exploration to rival any modern-day safari.

Jonathan Scott

Preface

SABERTOOTH CATS ARE AMONG THE MOST POPULAR OF PREHISTORIC animals, yet surprisingly little information about them is available for the curious layperson. One particular genus, *Smilodon*, has exerted an intense fascination since its discovery, and it has been featured in children's books, cartoons and films. But there were actually many other genera and species of sabertooth cats, coming in different shapes and sizes. To define them in a single sentence, sabertooth cats are extinct members of the extant cat family (Felidae), and thus close relatives of our living cats but different from them in several ways—most notably in having spectacularly enlarged upper canines, but also in a series of anatomical features that point to a different hunting style.

Sabertooth cats are not the only subject of this book, because they were not the only sabertooths to exist. As used by paleontologists, the term "sabertooth" designates also several kinds of extinct predators that were not cats, or even close relatives of them, but that shared some or all of the distinctive anatomical adaptations of sabertooth cats. Nimravids, barbourofelids, thylacosmilids—each of these obscure names designates a wholly different family of predators that developed remarkably similar morphologies. Some were no taller than a domestic cat, others were larger than a lion, and some would have looked weird indeed. This book intends to review that diversity of sabertooths, cats or otherwise.

For specialists, sabertooths have posed some of the most baffling enigmas of paleontology, and there is still much to learn. What did sabertooths look like? Some reconstructions depict them essentially as lions or tigers with oversized fangs, while others show them as bizarre creatures not resembling cats, or any other living carnivores for that matter. How did they use their spectacular canine teeth? There have been many hypotheses about the predatory habits of sabertooths, ranging from theories that they would be utterly unable to hunt and would have been scavengers exclusively to those that they were capable of hyperviolent stabbing. And why did they finally go extinct? Some experts thought that sabertooths were victims of an irreversible trend, in which their canines became so big and cumbersome over the generations that they ultimately caused the demise of the last species. Others believed that sabertooths specialized in hunting gigantic, thick-skinned prey, and that at the end of the last ice age when many of those monsters vanished, so did the sabertooths—leaving the world to the faster "normal" cats, which were better adapted to hunt fleet-footed prey like horses and antelope. We will probably never be able to answer these questions with total certainty. But over the last few decades a lot of exciting research has been carried out that certainly is taking us closer to the answers, and it has already suggested that all of the old hypotheses reviewed above were, in all likelihood, wrong.

Continuing studies will reveal ever more details about the evolution and biology of sabertooths, but unless we manage to master time travel, one thing will always remain true: they will be visible to us only as dry fossils or as reconstructions of one kind or another. As with all fossil species, a sort of "visual translation" is needed before we can see them as living creatures, and this process of restoration is also an essential part of the contents of this book. Reconstruction is an art and a science, and it requires a familiarity with anatomy and a fastidious attention to detail. Some people might believe that since we will never know with absolute accuracy what extinct animals looked like, a rough approximation should be enough. But we don't perceive living animals in an approximate way – at least when we are at all interested in them: horse or dog enthusiasts, for example, know how important precise morphology is in order to define their favorite breeds; and wild animals look beautiful to us because of their unique, precise shapes and proportions. Subtlety is all-important when picturing a living creature, as wildlife artists have known since the times of the Altamira or Chauvet cave paintings. All fossil animals are one step removed from direct observation, so it is essential that we strive for total accuracy in the process of reconstruction, even though – or precisely because – it is an ultimately unattainable goal. In the case of sabertooth cats, careful attention to anatomical detail is the only way to get a realistic idea (one not based on subjective preconceptions) of how similar or how different they were from their modern relatives, the extant big cats.

A strong commitment to science has been the hallmark of good "paleoart" since its beginnings, and many paleontologists also have a great capacity for visual observation and considerable drawing abilities. The founding father of vertebrate paleontology, G. Cuvier (1769–1832), was a competent draftsman, and around the beginning of the nineteenth century he produced remarkable restorations of some of the fossil species that he described for science. Unfortunately, he did not publish those drawings, which remained forgotten in the archives of the Muséum national d'Histoire naturelle in Paris until they were rediscovered in the late twentieth century (Rudwick 1992). If Cuvier had been less modest (or cautious) about his artwork, maybe modern paleoart would have matured decades earlier than it did, but almost a century had to pass before the greatest pioneer of this discipline brought it to fruition: C. Knight. Besides his solid training as an artist (and an abundant talent), Knight was an accomplished anatomist and a keen observer of nature, and his collaboration with the great paleontologists of his time – especially H. F. Osborn – was a process of constant discovery (Milner 2012). More to the point of this book's subject matter, Knight's collaboration with paleontologists led him to create restorations of sabertooths that have stood the test of time in a remarkable way, thanks to the rigor and beautiful simplicity of his working methodology.

In this book, I hope the reconstructions will serve as both a tool of study and a way to make available to readers the diversity and depth of anatomical detail that the fossil record of sabertooths has revealed to scientists after many decades of research. Fossils are recovered only through strenuous fieldwork, followed by patient cleaning and preparation of the specimens and endless hours of analysis, measurement, and the processing of CT scan images, or whatever technique is used to extract information from the fossilized remains. Should the knowledge so laboriously amassed remain buried in academic publications, beyond the reach of sincerely curious lay readers? I don't think so.

Beyond the interest that any group of fossil animals may have for both specialists and laypersons, the study of extinct creatures and their adaptations gives us a renewed appreciation of how nature works. Furthermore, paleontology provides a perception of the temporal dimension of life that today is more necessary than ever. In our ancestors' distant past, when hominids were just one more kind of mammals completely subject to the laws of ecology, the perception of nature from their own spatial and temporal scale was all that our ancestors required to meet their everyday needs. Even today this perception (an "immersion" in nature) greatly increases our well-being. But, having become collectively the most powerful biological agent on earth (still subject to the laws of ecology but in a way that the individual human, sheltered by technology, easily fails to notice), we badly need the opposite view as well: to see our home planet with perspective, to perceive its fragility and realize how crucial our present actions will be for the long-term future of the biosphere – and of our species.

Seeing the earth from space has provided such a perspective in spatial terms, and paleontology does something comparable in temporal terms, showing us that extant biodiversity, with all its fascinating detail, is just one frame in the long film of the evolution of life. Each species, which may seem static to us, is actually the result of the accumulated changes of countless species before, and it may be destined to keep changing or to go extinct, but that fate should not depend on human actions. It is deeply unethical for us to consider cutting through that evolutionary process without remorse as if our species collectively were a blind, impersonal agent of change, something comparable to the asteroid that killed off the dinosaurs 65 million years ago.

The long history of coevolution, as exemplified by the mutual influence between predator and prey through time, stresses the interdependence of all parts of the global ecosystem. We must ensure that the delicate machinery of the biosphere will keep functioning for the coming generations (if only to provide a habitable world for them), and that life will continue to evolve in the long term as close as possible to the way it would have done without our intervention. In a truly sustainable future, big cats and other carnivores, not human pressures, should be the main agents checking the populations of wild herbivores, wherever those are left. Predation is not only a drama to add emotion to nature documentaries: like it or not, it is an essential part of the way the web of life works.

Like modern big cats, sabertooths are iconic creatures, and this makes them excellent "ambassadors" of past biodiversity, as the emotional response they stir in us helps to heighten our interest in the details of their adaptations. Indeed, with their development of strikingly similar morphologies in so many independent lineages, they offer one of the best and most engaging examples of the laws of evolution in action. Additionally, there is a final, almost tragic quality to their extinction that helps us understand how wonderful it is that whenever we want to know more about the behavior and adaptations of lions or tigers, all we have to do is go where they live and observe them (at least, that will remain a possibility for a while). In contrast, we can throw all our scientific tools at sabertooths, yet there is an enormous amount about them that we will never know for sure. Nevertheless, in the process of studying them, we can learn much about the workings of predation in general, and our approach to the conservation of today's top predators should benefit from such insights.

Biodiversity is subject to time, and it thus changes, but it feeds our minds in a way that nothing else can. In fact, we should defend it not only as a material resource but also as the source of our sanity. Science and art, now more than ever, should naturally celebrate that diversity, and sabertooths are particularly fascinating examples of it.

Acknowledgments

MY INTEREST IN SABERTOOTHS BEGAN WHEN I WAS EIGHT YEARS OLD and lived in Valladolid, in Spain. I came across a copy of *The Golden Book Encyclopedia of Natural Science*, and leafing through the section on fossils, I found an illustration by Rudolph Zallinger that depicted a scene in the Pleistocene of Rancho la Brea, in California. There, a sabertooth was seen attacking a mammoth that had been trapped in the asphalt, while dire wolves and giant condors approached from the distance. That image had a striking effect on me: rather than just looking at an illustration, it was like being sucked through a time portal and taken back to a legendary era crowded with powerful beasts. And most important, this was no imaginary land with fantastic creatures, but a depiction of animals and landscapes that really existed. So I must credit the artwork of Zallinger for planting the seed of a fascination that lasts to this day.

Years later, as a teenager living in Caracas, I found in the school library a copy of *The Land and Wildlife of South America*, which included a chapter about extinct mammals illustrated by Jay Matternes. The paintings were fascinating in themselves, but what impressed me most was a double-page spread with a collection of sketches showing how Matternes had created his reconstruction of the sabertooth marsupial *Thylacosmilus*. This was a revelation as important as Zallinger's La Brea scene. Here I found not only information about a kind of sabertooth whose existence I had not suspected, but also a description of a whole process of acquisition and interpretation of scientific information in order to depict an extinct animal. I owe to Matternes the conviction that began to take shape in me around that time: that was the kind of work I would like to do if I could.

In spite of that fascination with paleontological subjects, I chose to get a formal artistic training, but later on I found that I needed to learn much more about paleontology in order to provide a solid base for my reconstructions of past life. That need opened a whole new chapter of learning, and I have been fortunate to meet a lot of people who have provided invaluable help along the way.

I first met Alan Turner in 1991, when I was a hopeful young paleo-artist looking for advice and collaboration. Alan was visiting Madrid on his way to a scientific conference in northern Spain, and he was invited to give a lecture at the Museo de Ciencias Naturales. I quickly put together a portfolio of my reconstructions to show him at the end of the talk, and to my great happiness he was instantly supportive of my work. During the few days of his stay we met several times, and we agreed to start work on our first collaboration, which would become the book *The Big Cats and their Fossil Relatives* – in several ways a senior companion to this volume. Many academic articles and books done together would follow, as well as a great friendship. We shared not only scientific discussions, but also family vacations and even flamenco evenings. This book would have been simply impossible without the long years of collaboration with Alan, from whom I learnt much about paleontology, about communicating science, and about many other things – including how a sense of humor can be the best and sometimes the only way to coexist with the contradictions of life. Alan passed away untimely, in early 2012, but he remains very much alive in the memory of his many friends and collaborators.

Also in 1991 I met Richard Tedford at the American Museum of Natural History, and starting with that first visit and in the years that followed he was always a source of help and inspiration. He opened for me that treasure trove of sabertooth fossils that are the collections of vertebrate paleontology of the American Museum of Natural History, and he freely shared his wisdom about carnivore evolution. All who knew Tedford agree that in addition to all his qualities as a scientist, he was the epitome of a gentleman.

In the early 1990s I started a fruitful correspondence with the Swiss paleontologist Gérard de Beaumont, who was already retired but more than willing to share his views about sabertooth evolution in a series of long manuscript letters that I now treasure (that was before I entered the e-mail era, although it may now be difficult to imagine that such a time ever existed). I count myself fortunate to have shared his knowledge in his late years. Also in the early 1990s I got in touch with the French paleontologist Leonard Ginsburg, who shared not only his knowledge but also the amazing fossil material housed at the Muséum national d'Histoire naturelle, in Paris, including some then-unpublished specimens that helped me get a clearer picture of early felid evolution. Leonard was a monumental figure in French paleontology, and his scientific legacy is enormous.

Also in the early 1990s Roland Ballesio from the University of Lyon, the author of a groundbreaking monograph about the European scimitar-tooth cat *Homotherium* and a leading expert in carnivore evolution, was kind enough to share many insights into the adaptations of sabertooths.

My longest collaboration in the study of carnivore evolution is with Jorge Morales from the Museo de Ciencias Naturales, in Madrid, a collaboration that continues to this day. Back in 1987 I approached him with a great fascination about fossil carnivores and an even greater ignorance of the subject, but he encouraged me to visit his laboratory whenever I wanted and virtually opened up the world of paleontology for me. He supported both my interest in paleoart, getting me in touch with patrons who would provide me with professional assignments, and in research, giving me the fundamentals of vertebrate paleontology and showing me the way into the academic literature. Later, when the incredible fossil site of Cerro Batallones, the star of European sabertooth paleontology, was discovered near Madrid, Jorge invited me to be there from the beginning and to experience firsthand the discovery of the best fossil sample ever found of Miocene sabertooths.

Manuel Salesa, who under Jorge's direction wrote his PhD dissertation on the anatomy and evolution of the sabertooth felid *Promegantereon* from Batallones, has become my closest collaborator in the study of sabertooths. It has been my privilege to share Manuel's research since his student years up to the present day, when he has established himself as one of the leading authorities on the subject of carnivore evolution. Together with Alan and Jorge, we have enjoyed the experience of making some really exciting discoveries in carnivore paleontology. A tireless field paleontologist, Manuel has excavated many of the fossils that have changed our view of sabertooth evolution over the last few years. He has helped me not only with discussions of many scientific topics for this book, but also in more down-to-earth ways, even with the labeling of many illustrations. He bears with me with the patience and enthusiasm that only real friends can show, and he is the best companion one could wish for in this continuing quest for the sabertooths.

Back in the mid-1990s, Angel Galobart from the Institut Catalá de Paleontología, in Sabadell, invited me to collaborate in the study of the fossils of *Homotherium* from the Pleistocene site of Incarcal, in Gerona. That was the start of a satisfying collaboration, which has resulted in several joint papers and an enduring friendship.

One essential aspect of my work in the reconstruction of sabertooths is the research into the anatomy of extant big cats, and that implies the dissection of actual specimens. In this task I have had the enormous help of the anatomist Juan Francisco "Paco" Pastor from the University of Valladolid. Paco's enthusiasm for anatomy is outstanding, and over the years he has created an amazing collection of comparative osteology that has been an essential reference for many paleontology students. He has organized many dissections of big cats for us, which have provided a solid anatomical ground for my reconstruction work and have also been the basis for several joint academic papers.

Over the years I have enjoyed very useful discussions with many paleontologists, including Jordi Agustí, John Babiarz, Jose María Bermudez de Castro, Meave Leakey, Larry Martin, Stephane Peigné, Blaire Van Valkenburgh, Lars Werdelin, Stephen Wroe, and Xiaoming Wang. Museum specialists and curators who have kindly provided access to specimens include Francis Thackeray (Transvaal Museum, Pretoria), Jerry Hooker (British Museum, Natural History), Abel Prieur (Université Claude Bernard, Lyon), Michel Philippe (Lyon Museum d'Histoire Naturelle, Lyon), and Deng Tao (IVPP, Beijing).

Gema Siliceo has painstakingly processed many computed tomography images that have been essential for our study of both fossil and extant carnivores, and she provided the images for figure 4.8, as well as helping with the labeling of several other figures. The CAT images themselves were obtained at the Hospital Universitario de Valladolid.

Oscar Sanisidro kindly helped with several figures for chapter 1, in particular figures 1.2, 1.4, 1.5, 1.7, and 1.8.

David Lordkepanitze provided useful discussions about the saber-tooth-human interactions at Dmanisi, and he also provided a nice cast of a *Megantereon* skull from that site.

One important facet of my work has been the opportunity to learn about 3D computer modeling and animation and its applications to reconstruction. In 2004 I started collaborating with the computer animator Juan Perez-Fajardo, who has put enormous enthusiasm and effort into our joint projects. I will never see animal action and locomotion the same way after this experience.

Jonathan Scott has been kind enough to write a foreword for this book, providing a complementary vision from his privileged standpoint as a field zoologist and one of the people with a greatest firsthand knowledge of the behavior of big cats in the African wilderness. His lively presentation of the BBC series *Big Cat Diary* has given millions an intimate view of the lives of African felines, and to me in particular it has provided inspiration for the task of trying to put together the data that will really bring the sabertooths back to life, not as mere scientific hypotheses but as the breathing creatures they once were.

For the task of picturing the big cats as animals of real flesh, my time spent in the African wildlife reserves has been vital, and the highly professional work of the guides and drivers there has made all the difference. They are a special breed of field naturalists, and the best among them have an unparalleled knowledge of animal behavior. Andrew Kingori in particular has made my recent safari experiences into true lessons of big cat natural history.

Pedro Bigeriego has been a great companion (and driver) on safari, and he is a source of constant support in this and other projects.

For much of the last two decades, the Museo Nacional de Ciencias Naturales, in Madrid, has been my scientific home, allowing me to conduct my research and enjoy continuing collaborations with the fabulous team in the Paleobiology Department. Since 1991 the government of the Comunidad Autónoma de Madrid has provided funding for the excavations at the Cerro Batallones fossil site, which have been the source of the raw material for many of our findings about the early evolution of sabertooth felids.

At Indiana University Press, James Farlow, editor of the Life of the Past series, has been enormously supportive from the start, and so has Robert Sloan, the science editor. Chandra Mevis and Michelle Sybert managed the manuscript, and Jeanne Ferris patiently copyedited the text, painstakingly spotting any inconsistencies.

My wife, Puri, and son, Miguel, bear with me through this and all my other crazy projects, to which I usually devote more time than would seem to fit the concept of a good parent or husband. And what is worse, they know that I will keep demanding their patience in the foreseeable future. I am really fortunate to have them.

My parents, Severi and Florencio, realized early on that I would not be pursuing a conventional career, and at all times they supported my unlikely professional choices, as has my sister, Maite. I know it has not always been easy, and I will always be grateful.

Many other kind people have been of help during the long process of developing this project. To all of them goes my most sincere gratitude. This book is in many ways a dream come true, and it would have been impossible to make it real without their help.

Sabertooth

1.1. Size comparison of the Pleistocene felid sabertooth *Smilodon populator* (background) and the Eocene creodont sabertooth *Machaeroides eothen*.

What Is a Sabertooth?

TODAY SABERTOOTHS ARE A FAMILIAR KIND OF EXTINCT CREATURES for scientists and for many laypersons, but in the early days of paleontology, not even scientists knew that such a thing as a sabertoothed predator had ever existed. Consequently, when early paleontologists first tried to make sense of fragmentary fossils of sabertooths, they attributed the remains to other, already known groups of animals. After all, those early discoveries were not of complete skeletons or even skulls, which would have revealed right away that the bizarre canines of sabertooth cats fit into an otherwise catlike skull and skeleton. Instead, the partial finds were like pieces of a jigsaw puzzle with no complete picture to refer to.

One of the first people faced with the task of interpreting sabertooth fossils was the nineteenth-century Danish naturalist P. Lund. In the 1830s, Lund devoted a lot of time and effort to exploring the caves of Lagoa Santa, in the Brazilian region of Minas Gerais. He had left Denmark in 1833 (at the age of thirty-two) to pursue botanical research in Brazil, but in 1834 he met his compatriot P. Claussen, a fossil collector who had worked in Argentina before coming to Brazil. Lund was immediately fascinated by fossils and paleontology, so he abandoned botany and moved to Lagoa Santa, then a village with fewer than five hundred inhabitants that was surrounded by numerous calcareous caverns, some of them rich in fossils. Many of the caves were actively exploited for saltpeter, with lots of fossils being destroyed in the process, so Lund set out to salvage as much material as possible (Paula Couto 1955; Cartelle 1994). With admirable dedication, he and his local assistants explored cave after cave, collecting over 12,000 fossils between 1834 and 1846. It was extremely hard work, but Lund's fascination for the extinct fauna of Brazil led him through all the difficulties he encountered, and whenever he found a rich fossil site, his imagination was set aflame. One of his peak achievements was to find the first remains of perhaps the most spectacular sabertooth cat, the Pleistocene felid *Smilodon populator*. Lund's first finds were just a few isolated pieces, and he thought they belonged to a hyena, naming the creature *Hyaena neogaea* in 1839. But in 1842, with the addition of a little more material, including a few more teeth and some foot bones, Lund–an adept follower of G. Cuvier, the father of comparative anatomy–soon realized that the predator actually belonged to the cat family.

Lund was convinced that the numerous bones of large mammals that he found in the caves had been dragged there by big predators, which retreated to their dark dens to feed at leisure. The identification of *Smilodon* as the dominant predator of its time rounded out the scenario in his mind:

Discovering Sabertooths

Regarding its size, this unique extinct carnivore rivaled the largest known cats or bears; the size of its canines is very much larger than in any species of carnivore, living or fossil. Judging by the dimensions of its foot bones, its body must have been heavier than that of any of the living felines, including the lion.

It is evident that a carnivore of such size, armed with such formidable weapons, must have reaped abundant victims among the inhabitants of the ancient world. In fact, I found the remains of its prey in three different caverns, which included, without exception, great accumulations of bones of diverse animals, many of them of gigantic size . . .

In view of the unusual form of the canines of this animal, I propose for its generic designation the name *Smilodon* ("tooth shaped like double-edged knife"). Its bloody deeds, whose memory still endures in the caves that served it as dens, doubtlessly qualify it for the specific name of *populator*, "he who brings devastation." (quoted in Paula Couto 1955:7–8)

Lund's account is a fitting introduction for an impressive animal that no human being had seen in over ten thousand years, and that no one would ever see alive again. While his interpretation of the origin of the fossil accumulations in the Lagoa Santa caves as the result of the activities of *Smilodon* is now thought to be not quite correct, his assessment of the size and strength of the newly discovered creature was soon confirmed by the appearance of more complete remains. In 1846 he could proudly write: "I now possess nearly all parts of the skeleton of this remarkable animal of the prehistoric world, mostly from different individuals" (quoted in Paula Couto 1955:7–8). During the few years following the initial description of *Smilodon*, the pace of discoveries quickened spectacularly. In 1843 a complete, amazingly well-preserved skeleton was found near Lujan, in Argentina, by the naturalist Francisco Javier Muñiz, and it was sent to the Museo de Ciencias of Buenos Aires. Around the same time, a complete skull was found by Claussen, Lund's friend, in a cave in Lagoa Santa, and sold for 2,000 francs to the Académie des Sciences de France, which in turn donated it to the Muséum national d'Histoire naturelle of Paris, where it can still be admired. A second, beautiful skeleton was found in the 1870s by a M. Larroque near Buenos Aires; it was obtained by the American Museum of Natural History of New York, where it remains on exhibit. The American paleontologist E. D. Cope published several drawings of parts of this skeleton in 1880, but the specimen remains mostly undescribed.

In the second half of the nineteenth century, remains of *Smilodon* from the Pleistocene were also discovered in North America, although the fossils there were scantier–but that situation would change. In 1875 Major H. Hancock, then owner of the Rancho la Brea property in Los Angeles County, presented a *Smilodon* canine tooth, recovered from the asphalt pits on his property, to Professor W. Denton of the Boston Society of Natural History. In spite of the impression that this gift made, it took twenty-five more years for that society to join forces with other institutions and organize the first scientific investigations at La Brea. Then, during the first decade of the twentieth century, an enormous amount of fossil mammal bones were recovered from the site (Harris and Jefferson 1985). Among these were literally thousands of bones of *Smilodon*, representing hundreds of individuals of the North American species S. *fatalis*, closely related to but distinct from its South American cousin.

Thus in the early twentieth century, *Smilodon* was a well-established element of the known prehistoric world. It was not the first sabertooth to be known to science – that privilege apparently belongs to the genus *Megantereon*, ironically mistaken for a bear by Cuvier. But *Smilodon* was the first genus to be known from reasonably complete fossils; it remained the best-known one for many decades; and it still includes the biggest and most spectacular species recorded.

So What Is a Sabertooth?

Smilodon is such a spectacular fossil that it has become quite popular, and it was even named the state fossil of California. It is also one of the few fossil mammals to appear repeatedly in cartoons and movies, and as a result a popular image of sabertooths has taken shape: they are perceived as big cats, armed with impressively long fangs (the sabers), and they are often seen in the vicinity of cavemen and mammoths. This image has some truth in it, but it is certainly not the whole truth. Like many sabertooths, *Smilodon* differed from modern big felids in being very robust, with stocky and enormously muscular limbs, a long and strong neck, and a short tail like that of a lynx or bobcat. A study of its skeleton reveals many other, more subtle differences that, taken as a whole, point to a hunting style differing in several ways from that of modern cats.

The identification of the concept "sabertooth" with *Smilodon* has inevitably masked one of the main facts of sabertooth history: diversity. This book deals with sabertoothed predators, a broader concept than that of the sabertoothed cats. While the latter indeed belonged to the cat family (including animals like *Smilodon* itself and closely related genera such as *Megantereon* and *Homotherium*), many sabertooths did not, and some of them would look quite uncatlike to a modern observer. Some really were big and heavier than any living cat, but others were smallish creatures, scarcely larger than your average house cat (figure 1.1). Some were contemporary with early humans and mammoths, but much of the evolutionary history of sabertooths took place long before mammoths, humans, or even our earliest hominid ancestors existed.

How, then, can we define the sabertooths? Briefly, we can say that they were a group of fossil vertebrates, most – but not all – of them mammals; they were all predators; and they all possessed a set of anatomical features that define them as sabertooths, including the presence of elongated upper canine teeth and other adaptations in the skull that allowed them to open their jaws in the huge gapes necessary to bite with such enormous teeth. But this is a rather complex definition, and its various parts need to be discussed in more detail.

First of all, sabertooths are fossils. All of them became extinct so long ago that no human being ever saw one alive in historical times – to

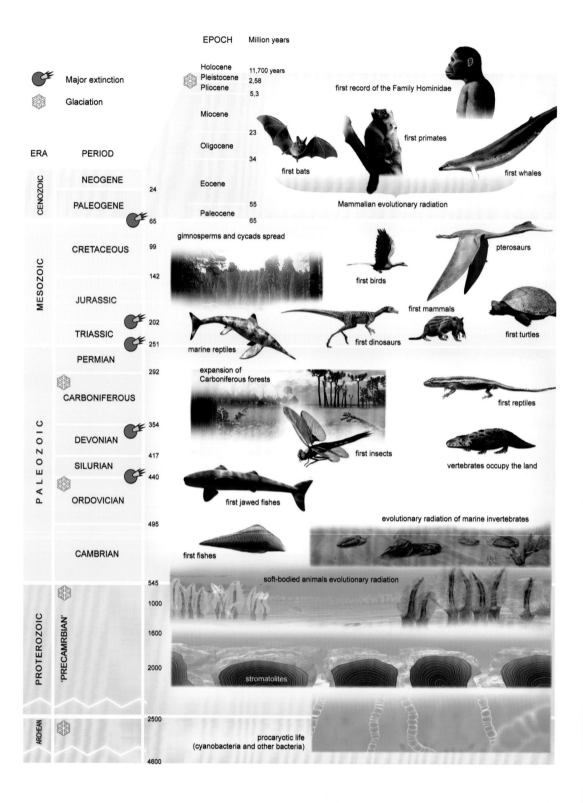

EPOCH | Million years

Major extinction
Glaciation

Holocene — 11,700 years
Pleistocene — 2,58
Pliocene — 5,3

first record of the Family Hominidae

Miocene

23

Oligocene

34

first primates

first bats

first whales

ERA | PERIOD

Eocene

24

CENOZOIC

NEOGENE

PALEOGENE

Paleocene

55

65

Mammalian evolutionary radiation

gimnosperms and cycads spread

CRETACEOUS — 99

142

pterosaurs

first birds

MESOZOIC

JURASSIC

first mammals

TRIASSIC — 202
251

first dinosaurs

first turtles

marine reptiles

PERMIAN — 292

expansion of
Carboniferous forests

CARBONIFEROUS

first reptiles

DEVONIAN — 354

first insects

417

SILURIAN

vertebrates occupy the land

ORDOVICIAN — 440

PALEOZOIC

first jawed fishes

495

evolutionary radiation of marine invertebrates

CAMBRIAN

first fishes

545

soft-bodied animals evolutionary radiation

PROTEROZOIC

'PRECAMBRIAN'

1000

1600

2000

stromatolites

2500

ARCHEAN

procaryotic life
(cyanobacteria and other bacteria)

4600

our knowledge, at least – so everything we know about them is based on their fossilized remains. Claims of modern sightings of sabertooth cats in South America and Africa remain completely unsubstantiated. And although some zoologists have suggested that the living clouded leopard of South Asia qualifies as an incipient form of sabertooth because of its long canines, the similarities between sabertooths and this extant felid are limited and rather superficial.

The history of sabertooths spans an enormous length of geological time. Although the latest species, such as the famous *Smilodon fatalis* from Rancho la Brea, disappeared "only" ten thousand years ago, the earliest mammalian sabertooths lived some 50 million years ago (or Ma) in the Eocene of North America. Even earlier, the gorgonopsians, a group of so-called mammal-like reptiles with many sabertooth features, lived in the Permian period, long before true mammals or even dinosaurs evolved, thus stretching the history of sabertooths, in the broad sense, as far back as 250 Ma (figure 1.2).

Second, all sabertooths were synapsids – that is, they belonged to the large group of vertebrates that includes both the mammals and the mammal-like reptiles. In anatomical terms, the skull of synapsids is characterized by the possession of a single cranial aperture behind the orbits (or eye sockets), hence their name, which means "a single opening." In mammals, that opening corresponds to the large area in the side of the skull where the temporalis muscle attaches to the parietal bone. In contrast, the diapsids – a large group of vertebrates that includes the dinosaurs, crocodiles, and birds – originally had two openings behind the orbits. The more familiar-looking of sabertooths were all mammals, and during the last 50 million years there were quite a lot of species of sabertooths, belonging to several unrelated families, orders, and even infraclasses of mammals. Sabertoothed predators include the following groups:

1. The gorgonopsians, an extinct group of Permian therapsids, or mammal-like reptiles.
2. Thylacosmilid sabertooths, an extinct family of South American predaceous marsupials.
3. Machaeroidine sabertooths, members of the Creodonta, an extinct order of meat-eating placental mammals related to but different from the true carnivorans (members of the order Carnivora).
4. Nimravid sabertooths, members of the extinct carnivoran family Nimravidae.
5. Barbourofelid sabertooths, members of the extinct carnivoran family Barbourofelidae.
6. Felid sabertooths, or true sabertooth cats, extinct members of the living carnivoran family Felidae.

The last three families were members of the Feliformia, a major division of the order Carnivora that includes the modern cats, civets, mongooses, and hyenas.

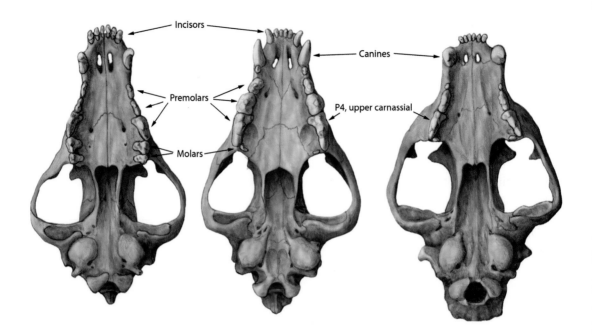

1.3. Skulls of modern carnivores with tooth and skull terminology. From left to right: wolf (*Canis lupus*), spotted hyena (*Crocuta crocuta*), and leopard (*Panthera pardus*).

It is not impossible that nonsynapsid sabertooths could evolve, but it is improbable, and certainly we haven't found any in the fossil record. Carnivorous dinosaurs such as the allosaurs, as well as modern varanid lizards, have sharp, flattened and serrated teeth that resemble the namesake canines of sabertooths to a remarkable degree, but otherwise the animals are just too different to be regarded as sabertooths.

Third, all sabertooths were predators. Apart from their fearsome-looking upper fangs, the shape of the rest of the dentition of sabertooths shows that they fed almost exclusively on meat. They were "hypercarnivores," meaning that their dentitions were so specialized for cutting meat that they had lost most of their ability to deal with other food items, such as bone or vegetable matter. Many modern mammalian predators, including wolves and bears, are more versatile, with dentitions that are suited to crushing bones, crunching insects, and processing vegetable matter. These animals possess molar and premolar teeth with different shapes adapted to those functions, in addition to the scissor-like, meat-cutting teeth known as the carnassials. Other modern carnivores–including all of the cats and many members of the weasel family, or mustelids–are true hypercarnivores (figure 1.3). But sabertooths went one step further in their predatory specialization. Therapsid sabertooths were obviously predatory too, but–unlike other sabertooths–they lacked any substantial dentition behind their canines, and they would have swallowed large chunks of meat whole, in truly reptilian fashion.

Other groups of mammals have developed elongated upper canines, more or less like sabers in appearance, but their function is related to display, defense, intraspecific aggression, or the manipulation of food items, rather than the hunting of large prey. Therefore, these creatures–including the extant musk deer, chevrotains, and walruses, as well as the bizarre, vaguely rhino-like uintatheres of the American Eocene – do not qualify as sabertooths.

In the fourth place, all sabertooths share a series of morphological modifications in their skulls and skeletons. Of course, the most obvious feature is the elongation of their upper canines, and most of the other modifications are related to the use of the canines. These modifications, which can be defined collectively as the "sabertooth complex," affected articulations and muscle insertion areas in the mandible, skull, neck, back, and limbs, and they varied considerably among different saber-toothed animals. No two groups share all the features making up the sabertooth complex, but all have at least several of them in common, and in some cases the list of shared features is strikingly long.

One word that often occurs in the literature about the sabertooths is "machairodont," which means "knife-toothed" and derives from the same root as the genus name *Machairodus*, referring to a felid sabertooth of the Miocene. The term "machairodont" is often used in a general way for all things sabertooth and can be either a noun or an adjective, so what we call here "sabertooth features" can be also termed "machairodont features." However, the term "machairodontine" is more specifically applied to the sabertooth subfamily Machairodontinae within the cat family Felidae. The reader should be prepared to be patient with the occasional ambiguity of some scientific terms, which can easily confuse the layperson.

The fact that sabertooths were not a single group of animals, but a series of totally unrelated groups that developed similar adaptations independently, constitutes a dramatic example of a phenomenon known as convergent evolution. In spite of changing views about the precise mechanisms of evolution, there is a general consensus among specialists that the process starts within organisms, whose genomes are subject to more or less random genetic changes or mutations, and then continues as the environment allows individuals that carry beneficial or at least neutral changes to survive and reproduce, while individuals with maladaptive changes quickly die off.

This process, called natural selection, modifies organisms that are themselves constrained by the fact that their genomes determine a limited range of possible or probable changes (what we call phylogenetic constraint) – which makes it unlikely that two animals derived from very different ancestors will end up evolving into similar creatures. But this is precisely what happens in cases of convergent evolution, where the apparent similarities between two or more groups of animals are caused by their adaptation to similar functions, rather than by a close common ancestry. The environmental forces that favor the development of similar adaptations in originally different organisms are what we call adaptive pressure, and the more distantly related two groups of animals are, and the more different their ancestral morphologies, the stronger will be the

Convergent Evolution

adaptive pressure needed to produce convergent evolution. Thus, the degree of convergence between unrelated groups of sabertooths suggests that, although their adaptations may look bizarre to us, they must have provided their owners with important advantages for survival, otherwise they would not have been selected by evolution over and over again.

The various groups of sabertooths, which initially attract our attention because of their striking similarities, also retain many features that betray their different evolutionary origins. The skulls and especially the dentitions of creodonts and carnivorous marsupials are so different from those of true carnivores that it is relatively simple (at least for the specialist) to tell creodont and marsupial sabertooths from "true" sabertooth cats. However, the similarities between more closely related sabertooth groups can be so detailed as to lead specialists to mistaken interpretations of their affinities. For example, members of the extinct carnivoran family Nimravidae – now known to be different from, and only distantly related to, the true cat family Felidae – converged so closely with the latter that they were long considered to be just a subfamily of the felids. For decades they were also known as "paleofelids" (or "old cats"), as opposed to the true felids that were referred to as "neofelids" ("new cats"), a naming system that reflected the belief that nimravids were just an early branch of the cat family. Only a detailed study of apparently obscure traits of their anatomy, such as the ear region of the skull, revealed the nimravids' true affinities (Hunt 1987). Such detailed convergence is obviously facilitated by the fact that all members of the Feliformia share a common ancestor in the early Tertiary. The concept of convergent evolution is essential to the understanding of the sabertooth adaptations, and the concept's applications to the problem of sabertooth evolution have been many, and sometimes contradictory.

In order to fully understand the implications of convergence, and sabertooth evolution in general, we need to consider some aspects of mammalian and vertebrate classification.

Aspects of Nomenclature and Classification

When I said that all sabertooths were synapsids, I introduced a major group of vertebrates that may be unfamiliar to many readers. Traditionally, terrestrial vertebrates have been grouped by zoologists in classes, including the Amphibia, Aves, Reptilia, and Mammalia (the amphibians, birds, reptiles, and mammals). But according to modern classification, mammals are part of a larger "natural group" of vertebrates, the synapsids. This group happens to include mammals and some reptiles, such as the therapsids or mammal-like reptiles, and the pelycosaurs, such as the famous sail-back *Dimetrodon* of the Permian. But it certainly does not include all reptiles, so it is not strictly a category "above" that of the class. Why, then, should we apparently break the rules of scientific nomenclature and talk about groups of animals that don't conform to the classic boundaries of the vertebrate classes? And what is the meaning of the term "natural groups" in this context? To understand this, we first need to take a look at the origins of the formal classification of animals and consider what is happening to it today.

The rules of taxonomy, the classification of living beings, and their nomenclature, or naming, were devised in the eighteenth century by the Swedish biologist C. von Linné (better known by the Latinized name of C. Linnaeus), and are known as the Linnaean or binomial system. Each species is known by a pair of Latin or Latinized names, written in italics: first the genus name, such as *Homo* in our own case, and then the species name, *sapiens* in our case. This system includes each species in ever larger, more inclusive categories, so that we humans (*Homo sapiens*) belong in the family Hominidae, within the order Primates in the class Mammalia, itself a part of the phylum Chordata, within the kingdom Animalia. We can thus know what creature we are talking about in spite of the different common names that animals are given in different languages, and this system is especially important in the case of fossil species, because fossil animals normally lack common names.

There have been attempts to create ad hoc common names for fossil animals, and for felid sabertooths in particular. The Finnish scientist B. Kurtén, one of the most successful popularizers of mammal paleontology, was in favor of this procedure and created names like "Western Dirktooth" (for the sabertooth *Megantereon hesperus*), "Gracile Sabertooth" (for *Smilodon gracilis*), "Greater Scimitar-tooth" (for *Homotherium sainzelli*), and "Lesser Scimitar Tooth" (for *Homotherium latidens*). This was a brave attempt to spare the general reader the effort of getting familiar with – not to mention figuring out the pronunciation of – the admittedly forbidding Latinized names. But unlike the natural process of people giving common names to animals they see, this is a reversed, somewhat artificial procedure, and it is subject to the instability of the definition of species in the fossil record. In other words, a lion remains a lion in spite of the fact that, over the last few decades, scientists have changed its scientific name from *Felis leo* to *Leo leo* and then to the currently widespread *Panthera leo*, following successive revisions of felid systematics. But imagine that paleontologists conclude that there was only one species of the sabertooth cat genus *Homotherium* in the European Pleistocene after all, and that the species name *Homotherium sainzelli* was invalid, and wrongly given to some of the larger, probably male, specimens of *H. latidens*. Incidentally that seems to be the case, and it leaves us in an awkward situation regarding the common names created to differentiate two species that probably weren't separate after all. So, for better or worse, readers interested in knowing about sabertooths should resign themselves to getting used to the Latin names!

The Linnaean system has been used to categorize all known animals and other living beings into species, genera, families, and so on, and it has proved to be a most useful approach. But the classification of living beings today still largely reflects the early observations of classical naturalists about the apparent similarities between organisms. However, since the time of Darwin it has become evident that the classification of animals should reflect their common ancestry—which is often masked, rather than revealed, by apparent similarities. As more and more of those similarities are discovered to be the result of convergent evolution, we realize that many of the traditional groupings are artificial, grouping together animals that actually are not linked by shared ancestry.

A "natural group," thus, is one whose members do share a common ancestor, and the current emphasis on that concept reflects the striving of modern biologists for a perfect fit between the classification of living things and their inferred evolutionary relationships. Just as relatedness among people reflects shared ancestry, and we like to trace our family tree to see who wore our surnames in centuries past (what we call genealogy), so too the definition of relatedness in the animal kingdom, and among all other living things, should ideally reflect shared ancestry (what we call phylogeny). The fact that all formal groups of organisms should reflect a common ancestry is one of the postulates of cladism, the leading modern school of systematic classification. Cladists consider only "natural groups," or "clades," to be valid, and members of such groups share a common ancestor as defined by a "node," or branching in the evolutionary tree. Cladism comes into conflict with traditional classification in several ways. For example, some traditional groupings mix species of different ancestry, and cladists call them polyphyletic and dismiss them as confusing. But other groupings, while being essentially correct because they combine only animals that had a common ancestor, are incorrect in the way they separate those animals from others that also share a common ancestor with them and should properly be part of the same group, or clade. These are called paraphyletic groups. One good example of this is the group traditionally called "synapsid reptiles" (and the whole class Reptilia, for that matter).

The Synapsida as a group was initially created as a subclass within the Reptilia, in order to classify a series of paleozoic and early mesozoic reptiles united by the possession of a single opening, or "fenestra," in the skull behind each orbit (figure 1.4). These reptiles shared several anatomical features with mammals, and thus became known as mammal-like reptiles. Further study eventually confirmed that this group, the therapsids, indeed included the ancestors of true mammals. In this context the word "reptiles" defines only a grade, or, a "stage" of evolution, at which point some features have been developed but others have not. Specifically, the Paleozoic therapsids had not yet developed the diagnostic, or defining, features of mammals, such as the presence of a single dentary bone in the mandible, articulating with the temporal bone of the skull. But cladists do not accept grades as a criterion for classification, so they reject the concept of a group of "Synapsid reptiles" that excludes the mammals. They do accept mammals as a natural group and the therapsids as a larger natural group that includes the mammals plus the cynodonts, the gorgonopsians, and other groupings. The Synapsida, finally, is a still larger group, which includes the therapsids plus the pelycosaurs. In this context, the concept of a class Mammalia as a group of similar rank to the

1.4. Phylogeny of the tetrapods. The four branches at the bottom of the illustration correspond to the Synapsida.

Reptilia or reptiles becomes absurd, because in evolutionary terms, the mammals are just a branch of the therapsid reptiles. So, turning back to our broader definition of sabertooths, we still can say in informal terms that most sabertooths were mammals while the gorgonopsian pseudo-sabertooths were reptiles. But in strict cladistic terms we can comfortably say that all of them were synapsids.

Sabertooth Evolution

The independent evolution of sabertooth adaptations in many groups of predators at different times in geological history is such a striking phenomenon that the American paleontologist C. Janis called it "the sabertooth's repeat performances" (Janis 1994). At any given time there have been only one or at most three families of sabertoothed predators living on earth, so whenever all groups became extinct globally, one would expect that sabertooths would never be seen again. However, in each instance, and after several million years, a new group has popped up somewhere to try the sabertooth experiment once more.

The first predators to develop sabertooth features were the gorgonopsians. The body plan of these Permian therapsids reveals their essentially reptilian grade of evolution, and because they were so different from the true mammals, their anatomical adaptations to a sabertooth style of

1.5. Phylogeny of the mammals. Note that the marsupial sabertooths (Thylacosmilidae) are in the branch at the top of this phylogeny, while the placental sabertooth families are in the two branches at the bottom.

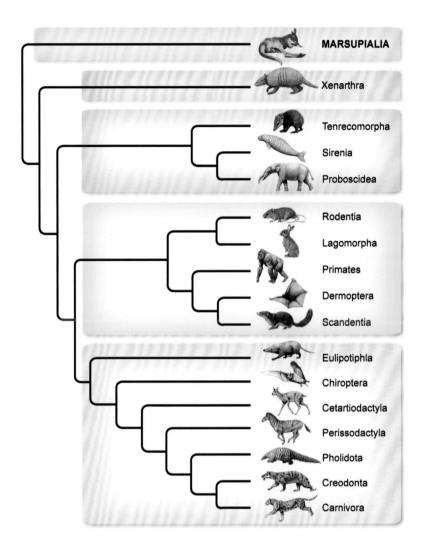

predation were very different from those of the mammalian sabertooths. Even so, gorgonopsians share with the latter the possession of very elongated, serrated upper canine teeth and a mandible that could open to very wide gapes. Like many but not all mammalian sabertooths, gorgonopsians also had large incisors arranged in an arc in front of the canine teeth. These animals disappeared at the end of the Permian.

Even if we focus our attention on the mammalian sabertooths, we still need to look at a very wide array of zoological groups. Most of them belonged to the infraclass Eutheria (placental mammals), but one group belonged to the infraclass Metatheria (marsupials), and its members were thus more closely related to kangaroos than to cats.

The thylacosmilids, or marsupial sabertooths (figure 1.5), were part of a larger group of native South American predaceous marsupials known as the superfamily Borhyaenoidea, which converged with placental carnivores in their general adaptations to eating meat (Goin and Pascual 1987). But unlike their placental counterparts, they never developed

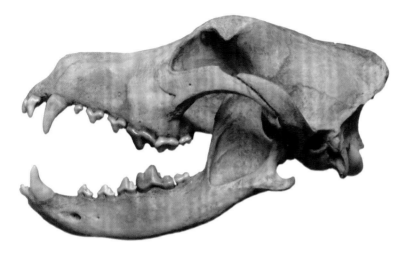

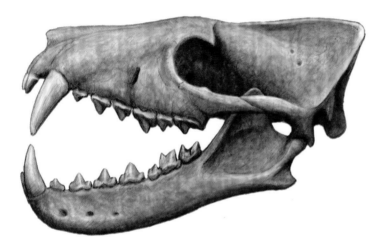

1.6. Skulls and mandibles of an extant canid, *Canis* (top), a member of the order Carnivora, and of the extinct hyaenodontid *Hyaenodon* (bottom), a member of the order Creodonta. The carnassial teeth are shown in red. Note that in the carnivores the carnassials are the upper fourth premolar and lower first molar, while in the creodonts they are one or two positions behind in the tooth row.

true carnassial teeth (a key dental adaptation of placental carnivores, as we shall see below), and the whole post-canine dentition of the thylacosmilids became somewhat blade-like and functioned like a very long meat-cutting device. The latest thylacosmilids were among the most specialized sabertooths, with enormously elongated upper canines that grew continuously, and a deeply modified skull showing extreme adaptations for a biting method far removed from that of other borhyaenoid marsupials and more similar (but not nearly identical) to that of placental sabertooths like *Smilodon*. Thylacosmilids disappeared in the Pliocene.

The next group of sabertooths, the machaeroidines, were creodonts, members of an entirely extinct order of predatory placental mammals called Creodonta, which evolved in the Paleocene along with the earliest "true" carnivorans (Dawson et al. 1986). The creodonts quickly diversified and evolved into larger forms in the early Eocene, while the true carnivorans remained small, weasel-like creatures for the whole Paleocene and much of the Eocene. Creodonts were related to the true carnivorans (they have been classified as a sister group of the Carnivora, "sister group" being

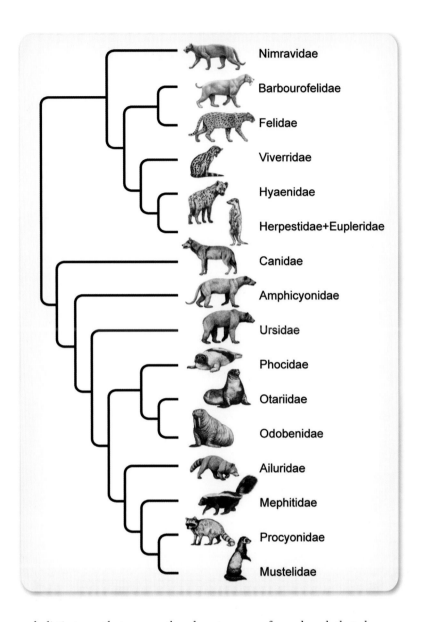

Nimravidae

Barbourofelidae

Felidae

Viverridae

Hyaenidae

Herpestidae+Eupleridae

Canidae

Amphicyonidae

Ursidae

Phocidae

Otariidae

Odobenidae

Ailuridae

Mephitidae

Procyonidae

Mustelidae

a cladistic term that means the closest group of equal rank that shares a common ancestor) and, like them, possessed a set of carnassials, specialized cheek teeth (also known as post-canine teeth because they come behind the canines) that became blade-like and thus well adapted to cutting meat. However, the carnassials of creodonts occupy different positions in the tooth row than those of the Carnivora (figure 1.6). Machaeroidines were not among the larger creodonts, ranging from the size of the domestic cat to that of the lynx. They disappeared in the middle Eocene.

Among the true carnivorans of the order Carnivora, the first group to evolve sabertooth adaptations was the family Nimravidae, which is first recorded in the Eocene, after the extinction of the machaeroidines. Members of the Carnivora are defined by, among other traits, the position of their carnassial teeth, which are the fourth upper premolar and the first

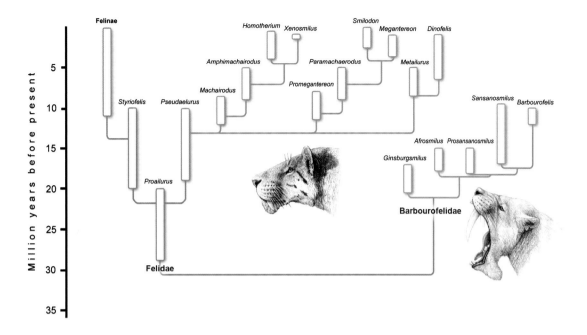

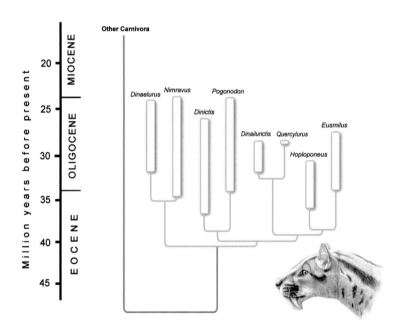

lower molar (figure 1.6). As mentioned above, nimravids were so similar to modern cats that they were classified just as another felid subfamily for many years (Bryant 1991, 1996b). But more detailed studies showed that their similarities to true cats were superficial and that the nimravids were a distinct family, not especially close to the cats but seen by some specialists as a sister group of all the other Feliformia. The nimravids disappeared near the Oligocene-Miocene boundary.

1.8. Phylogeny of the felids and barbourofelids (top) and the nimravids (bottom), showing the relationships among the most representative genera.

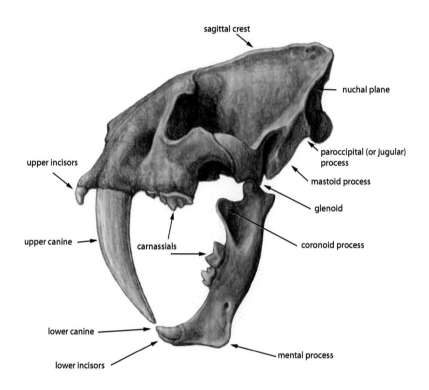

1.9. Skull of the felid saber-tooth *Smilodon* showing some of the osteological features discussed in the text.

sagittal crest

nuchal plane

paroccipital (or jugular) process

mastoid process

glenoid

coronoid process

upper incisors

upper canine

carnassials

lower canine

lower incisors

mental process

The Nimravidae were traditionally divided into two subfamilies, the more ancient Nimravinae from the Eocene and Oligocene, and the younger Barbourofelinae, from the Miocene. It is now clear that the differences between the two supposed subfamilies are too great to keep them lumped together in the same family, and a new family Barbourofelidae has been created (figure 1.7). In fact, it is possible that the barbourofelids were more closely related to true cats than the nimravids were (Morales et al. 2001; Morlo et al. 2004). Barbourofelids disappeared in the late Miocene.

The more familiar sabertooth cats belong in the carnivoran family Felidae, the "true cats" (figure 1.8). Felid sabertooths were in no way ancestral to modern cats. Rather, they were more like cousins to them, an independent group that for millions of years evolved side by side with the ancestors of our felines. They are usually thought to be a subfamily of their own, the Machairodontinae, while the modern cats are included in the subfamily Felinae. Some specialists further subdivide the modern, non-sabertoothed cats to include a subfamily Pantherinae for the big cats of the genera *Panthera* and *Neofelis*, and even a subfamily Acinonychinae, for the odd-looking cheetah. But recent molecular studies on the phylogeny of extant cats tend to give less importance to formal categories such as subfamilies and tribes, referring instead to "lineages," or groupings of genera that appear to cluster together because of their genetic affinities. In these classifications, the cheetah, in spite of its unique appearance, is clearly grouped with the puma and jaguarundi in a "puma lineage," and the species of *Panthera* and

Neofelis are also consistently grouped in a "panthera lineage" (Werdelin et al. 2010).

The close affinity of machairodontine sabertooths to extant felids is amply demonstrated by their anatomy, but the excellent preservation of some fossils of the most recent sabertooth species such as *Smilodon fatalis*, *Smilodon populator*, and *Homotherium serum* has made it possible to study their affinities via DNA analysis. In a recent study, R. Barnett and colleagues used refined techniques for ancient DNA extraction and analysis on samples of sabertooth DNA obtained from fossils of *Smilodon* and *Homotherium* found in North and South America, and the researchers also included in their study the extinct cheetah-like cat *Miracinonyx* from the American Pleistocene (Barnett et al. 2005). Their results confirmed the affinity of *Miracinonyx* to the extant puma-cheetah clade (and its especially close affinity to the puma), while indicating that the sabertooths were a sister group of the modern cats. Felid sabertooths went extinct at the end of the Pleistocene.

A Quick Look at Sabertooth Features

In order to get a general idea of what makes a sabertooth a sabertooth, it is useful to take a look at the non-sabertoothed relatives of the groups listed above and observe their broader differences. Within each group, we can see that the non-sabertoothed animals retain the more "primitive," or ancestral, features of their general groups, and it is their sabertoothed relatives that diverge more from the common ancestor—often showing a clear progression through time, at least when the fossil record provides us with the luxury of a phyletic series (that is, a sequence of related taxa arranged in time and showing what looks like an approximate relationship between ancestor and descendant). So, although it may sound strange, the modern cats are in a sense more "primitive" than their extinct, sabertoothed relatives, meaning that the former have changed less from their shared ancestors (figure 1.9). But what are the main differences between sabertooths and their non-sabertoothed relatives?

Let us start with the teeth. First of all, of course, there are the sabers (figure 1.10). These are the upper canines, which in sabertooths become long (or, in technical terms, high crowned), curved, and laterally flattened. The degree of elongation, curvature, and flattening is very variable, but overall the distinction between sabertooth and non-sabertooth canines is clear. In the best-known lineages, the earlier species have only slightly flattened canines that more closely resemble those of their non-sabertoothed ancestors, while the latest species have the most exaggerated saber shape. Many sabertooths display serrations (more technically known as crenulations) on the cutting edges of the sabers. Those serrations are not limited to the upper canines, and in some species all teeth are serrated when unworn. The serrations vary from very fine to rather coarse, and they make the teeth more effective at cutting through flesh and skin.

The impressive shape of sabertooth canines is just an exaggeration of the trend among synapsids to evolve teeth of different shapes in different

1.10. Skull comparisons of sabertoothed (left side) and non-sabertoothed (right side) borhyaenoids, nimravids, and felids. Top row: left, *Thylacosmilus,* right, *Borhyaena;* center row: left *Hoplophoneus,* right *Nimravus.* Bottom row: left *Smilodon,* right, *Panthera.*

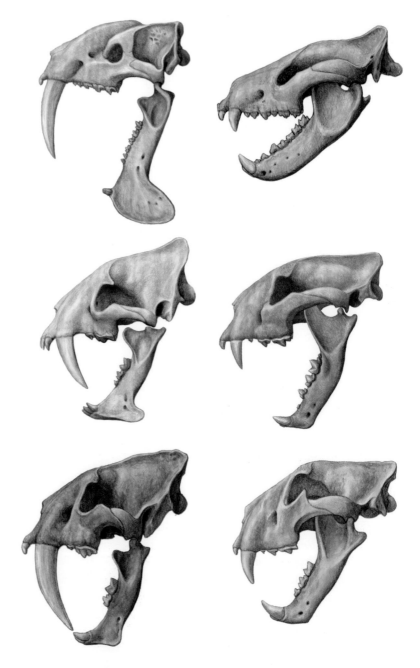

positions in the tooth row—a phenomenon called heterodoncy—in contrast to the relatively uniform tooth rows of most reptiles. Gorgonopsian sabertooths already displayed some degree of heterodoncy, and their upper canines were very long and had serrated borders, but were not nearly as flattened as those of the more specialized mammalian sabertooths (figure 1.11).

The rest of the teeth show their own set of differences. The incisors of sabertooths tend to be enlarged and arranged in an arc, while their lower canines often become smaller and are arranged as part of the battery of

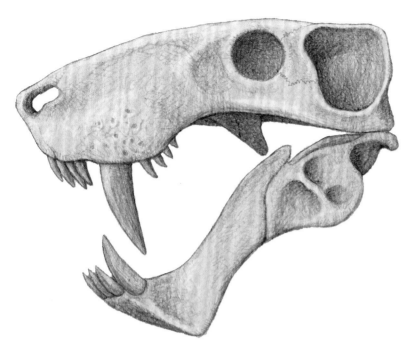

1.11. Skull of the gorgonopsian *Inostrancevia*. Note how the teeth have different shapes and sizes in different regions of the maxilla and mandible, a condition known as heterodoncy.

lower incisors. In the gorgonopsians, the incisors are more numerous than in mammals and are arranged in an impressive arc, but the lower canines remain large, although considerably smaller than the upper ones. In the marsupial thylacosmilids, the lower canines are much smaller, almost peg-like; but the incisors are smaller still, even vestigial. This is in striking contrast to the more typical sabertooth groups, which implies some mysterious functional difference. In some lineages of placental mammalian sabertooths, it is possible to trace how the lower canines, still large in the earlier species, get more and more reduced in later forms, while the lower incisors get larger and more protruding.

The cheek teeth tend to be reduced in both number and size in sabertooths. In gorgonopsians there were no substantial cheek teeth; in fact, these animals show no trace of the development of molariform (that is, resembling our own molar teeth in being broad and blunt) cheek teeth, a feature that would only appear in later, more mammal-like, therapsids. In marsupials, as we have said, there are no true carnassial teeth. And there is no great reduction of cheek teeth in thylacosmilids, where the whole row of cheek teeth works as a long cutting device, although it lacks the refinement of the true carnassial teeth of placental predators. In placental sabertooths the carnassials become quite elongated, with an exaggerated blade-like shape, but most other cheek teeth become smaller or disappear.

Regarding the general shape of the skull, in sabertooths it tends to be relatively high, with big crests for the insertion of the muscles that close the jaws (figure 1.9). Among the gorgonopsians the structure of mastication muscles had not attained the mammalian grade yet, and the mandible and skull were articulated via different bones (the quadrate

1.12. Full body reconstruction of the felid sabertooth *Homotherium* in four views. Note the long, vertically oriented legs.

and articular) than those involved in the mammalian articulation (the temporal and dentary). As we shall see in chapter 4, the gorgonopsians solved the problems of biting with large gapes in a rather peculiar manner, totally different from that of their mammalian counterparts. The glenoid process, which is where the articulation of the skull with the mandible is located in mammals, is usually placed more ventrally – that is, lower down – in sabertooths than in non-sabertooths. The mastoid area, which is the part of the skull situated just behind the ear opening, shows profound differences. In sabertooths, the mastoid process ("process" is the anatomical term for a bony protuberance, and the mastoid in particular is placed in the temporal bone of the skull, right behind the ear opening) is projected ventrally, partly enclosing the ear opening and almost touching the postglenoid in some cases, while the adjoining paroccipital process (a protuberance placed just behind the mastoid) becomes ever

1.13. Full body reconstruction of the marsupial sabertooth *Thylacosmilus* in four views. Note the short legs.

less projected and in some cases appears vestigial. The occipital plane, or the back of the skull, is often verticalized, in contrast with its more inclined orientation in non-sabertooths.

In the mandible of sabertooths, the most evident difference lies in the anterior part (the "rostral" part, in formal anatomical nomenclature), where the two halves or hemimandibles join (figure 1.9). That region is called the mandibular symphysis, and in sabertooths its anterior plane is high and vertically oriented, forming an angle of about 90 degrees with the horizontal plane of the mandible, in contrast with the gentle curvature of the symphyseal region in most non-sabertoothed predators. Additionally, the symphysis is strengthened, and its lateral margins are often prolonged ventrally to form a mental (or belonging to the chin) process of varying length. Another marked difference in the mandible is the reduction of the coronoid process (that is, the protuberance rising from the mandible in the space between the teeth and the articulation with the skull).

Behind the skull (figures 1.12, 1.13, and 1.14), the differences are very variable. In general, sabertooths tend to have long necks with enlarged

1.14. Full body reconstruction of the gorgonopsian sabertooth *Rubidgea* in four views. Note the flexed limbs and outwardly oriented elbows.

muscle insertions; robust backs that are often shortened and laterally stiffened; and very strong, robust forelimbs, capable of considerable lateral rotation. The limbs are relatively short in many (but not all) cases, especially the hind limbs and feet. The tail, where known, is often shortened (figure 1.15).

The precise functional meaning of these changes will be discussed in detail later, but at this point we can note how remarkably accurate is the repetition of many of these features in the unrelated groups of mammalian sabertooths.

Thus, we see that sabertooth adaptations have arisen time and again, creating a somewhat different way to be a predator. It is likely that we will discover many new sabertooth species in the fossil record; we probably already know most if not all of the major groups. Our knowledge has grown enormously since the first recognition of sabertooth fossils in the early nineteenth century, thanks to the systematic search for and excavation of fossil sites. Although many of the early

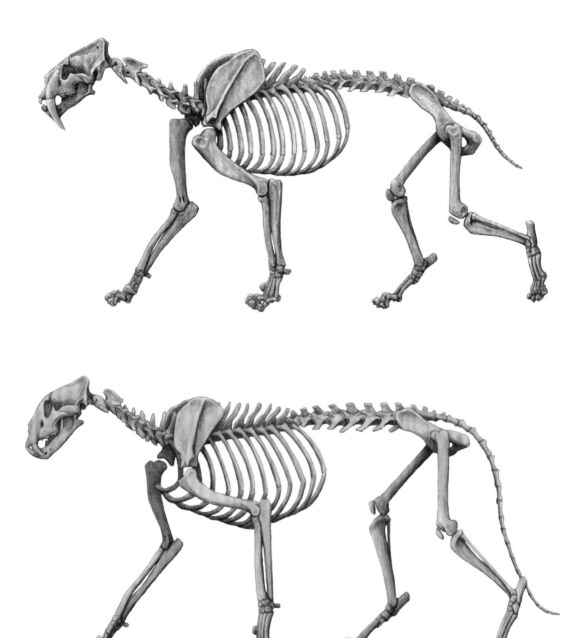

findings of fossils were a consequence of lucky chances, the current rate of discovery is based on a more precise knowledge of how and why fossils originate. Paleontology is a combination of biology and geology, and, as we shall see in the next chapter, it is only thanks to intimate knowledge of geological processes that paleontologists can understand where to look for fossils, and what the geological context of a fossil tells us about the circumstances in which the animals lived and died. Without such knowledge it would be impossible to get an accurate picture of sabertooths as living animals.

1.15. Skeleton comparison of two felids, the sabertoothed *Smilodon* (top) and the conical toothed *Panthera*. Note that the sabertooth has a longer neck, shorter back and tail, and more robust limb bones.

2.1. Schematic cutaway view of the Friesenhahn cave (Texas) to show the origin and preservation of the fossils. Top: the sabertooth *Homotherium serum* freshly dead in the cave. Center: the alluvial deposits are entering the cave and about to bury the skeleton. Bottom: the sediments have entombed the skeleton.

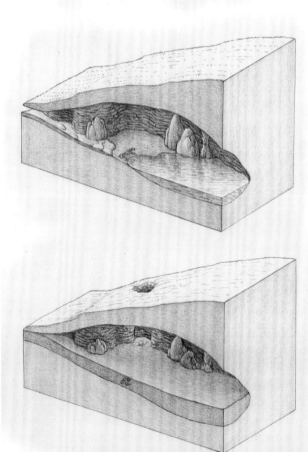

The Ecology of Sabertooths

2

DURING THE SPAN OF GEOLOGICAL TIME THAT SABERTOOTHS OF ONE kind or another have inhabited the earth, our planet has undergone dramatic changes. Continents have collided and then drifted away from each other; temperatures have oscillated wildly, from periods of scorching heat to chilling ice ages; sea levels have risen and fallen, changing the shape of coasts and alternately flooding and revealing thousands of square kilometers of land; and vegetation has changed, from otherworldly Paleozoic forests consisting of giant ferns and primitive conifers to Cenozoic communities made up of essentially modern plant types, but with distributions that fluctuated dramatically with climatic oscillations. The evolution of sabertooths has been tightly linked to these changes in their environments and to the evolution of other animal species, including their competitors and prey. Everything we know about their history has been gathered from a treasure trove of information encrypted in layer on layer of sedimentary rocks: the fossil record.

Sabertooths in a Changing Planet

Of all the millions of individuals belonging to all sabertooth species, only a handful (so to speak) have reached us as fossils, thanks to an improbable, almost miraculous chain of events: the fossilization process. Fossils are the remnants or traces—such as bones, teeth, eggs, leaves, roots, and footprints—of extinct living beings, produced after a set of physical and chemical processes called diagenesis results in the mineralization of the remains. What we usually call animal fossils are the creatures' hard parts (in the case of vertebrates, their bones or teeth), which, after being buried for thousands or millions of years, have suffered a more or less intense process of mineralization, as chemical substances percolating through the ground replaced the original organic tissues. There are other kinds of fossils, including the footprints (or ichnites) of extinct animals, which were preserved as the layers of mud on which the animals had moved hardened and got quickly buried under new sediment; or the rare imprints of soft tissue, preserved when the bodies of animals sank to the bottom of still, oxygen-poor waters where decay was slowed down and anaerobic bacteria took the place of the soft tissues, creating an outline of those tissues on the sediment. Some animals from the ice ages were preserved in the form of frozen carcasses buried in the Arctic permafrost, providing a wealth of information about soft anatomy and even external features, as exemplified by the woolly mammoth "mummies" occasionally found in the Siberian

Fossils and Fossil Sites

tundra. Nevertheless, these bodies did not pass through the diagenetic process, and they are just dead animals from the past.

Sabertooth enthusiasts keep hoping that one day the frozen body of an Ice Age *Homotherium* may be found in Arctic latitudes, but so far none has appeared. Nor has any sabertooth skeleton been found with traces of the animal's soft tissues around it; and we have had comparatively little luck with sabertooth footprints. So for the time being, most of the information we have about these animals comes from their bones. The preservation and discovery of vertebrate fossils is fraught with improbabilities, especially in the case of carnivores. For ecological reasons – after all, any animal needs to have a food source collectively weighing much more than it does – carnivores are much scarcer than herbivores in terrestrial ecosystems, making up no more than 2 percent of the total mammalian biomass. This imbalance is reflected in the fossil record, where carnivores make up only about 10 percent of all mammalian fossils found. At some exceptional sites called carnivore traps, the proportion is different or even reversed, and such sites are treasures indeed for paleontologists who study fossil carnivores.

There are different kinds of fossil sites where the remains of sabertooths can be found. Most of them are open-air sites that were formed as the sediments accumulating in the shores of lakes or rivers buried the remains of animals that had died in the vicinity. In such deposits, the regular pattern of accumulation of younger sediments on top of older ones allows scientists to determine the relative geological ages of the fossils preserved in them. Other sites correspond to cave deposits, where animal remains were similarly entombed in sediments, but the whole process in these cases took place within the limited space of natural cavities in the rocks. In the following sections we shall take a closer look at the different kinds of fossil sites.

Cave Deposits

Caves usually originate as water excavates tunnels and cavities in a matrix of limestone or dolomite; the resulting topography is known as karst. Cave floors often preserve sediments that have flooded in, sometimes incorporating the remains of animals. Cave fossils are usually difficult to date because there is no clear pattern of sedimentary succession in the deposits, so the age of an accumulation of fossils (sometimes called a "bone bed," if the fossils are found in high densities) is usually inferred by comparing the preserved faunas with those from well-dated open-air sites (discussed below). But cave sites have some advantages. For example, they often come from higher ground than typical valley-bottom accumulations, and thus they sample different environments and faunal associations. And because caves are often used as dens by carnivores, it is not unusual for their remains to be preserved there. These accumulations tend to be dominated by a single species, such as cave hyenas in several European Pleistocene cave deposits.

Friesenhahn Cave

One cavity that was clearly a sabertooth den is the Friesenhahn cave in Texas, dating to the late Pleistocene (Meade 1961). There, researchers found the remains of several individuals of the sabertooth felid *Homotherium serum* in association with hundreds of bones of proboscideans. Fossils of sabertooths of different ages, including cubs, were found at the site, reinforcing the notion that it was used as a den. The proboscidean bones, including those of mammoths and a few mastodons, are overwhelmingly those of young individuals that appear to have been the victims of the sabertooths, which would have dragged large portions of the carcasses back to the den to be eaten. One especially fine specimen is the skeleton of an old sabertooth, which was found articulated in a resting pose – suggesting that the animal simply lay down and died. Later, the cave floor was flooded with fresh sediment from the outside, burying the bones (figure 2.1).

Haile 21A

Another cave site that may have been a sabertooth den is the Irvingtonian (early Pleistocene, about 1 Ma) cave site of Haile 21A in Florida. This is the site where the first and only undisputed specimens of the robust homotherin *Xenosmilus hodsonae* were found. The partial skeletons of two individuals were discovered at this site, together with abundant remains of a peccary, *Platygonus cumberlandensis*, and some fragmentary fossils of *Smilodon gracilis* and of a gomphotherid mastodon. The authors who described the *Xenosmilus* fossils interpreted the site as a possible den for the sabertooths, and according to them these predators would have taken back to the cave most if not all of the peccary material (Martin et al. 2011).

Kromdraai Cave

Among the most famous cave sites are those in the Sterkfontein Valley in South Africa, notorious for their abundance of early hominid fossils from the Pliocene and Pleistocene and collectively known as the "Cradle of Humankind." Kromdraai cave in particular has yielded some of the best African fossils of the felid sabertooth *Megantereon*. The Kromdraai cave deposits are divided into "members," each corresponding to different accumulation episodes. Member A was interpreted by the paleontologist C. K. Brain (1981) as a carnivore lair, where sabertooths, leopards, and several hyena species brought the carcasses of their prey for quiet consumption. Member B was interpreted by E. S. Vrba (1981) as an occasional shelter for primates, but also as a death trap for some animals that fell down the steep opening shaft, and a source of opportunistic feeding for carnivores such as *Megantereon*. At least two adult sabertooths died there, so that large portions of their two skeletons were preserved side by side.

Zhoukoudian

The Pliocene and Pleistocene deposits in the cave system at Zhouk-oudian, China, became world famous because of the presence of homi-nid fossils, but there was also a den for carnivores, and remains of the Plio-Pleistocene felid sabertooths *Homotherium* and *Megantereon* have been found there. The main cavity at Zhoukoudian, known as Locality 1, yielded the hominid remains, and it has traditionally been interpreted as a place of intensive human occupation. The fossils of *Homotherium* were found in other areas, specifically Localities 9 and 13, but the skull of *Megantereon* is from Locality 1, as are a huge number of fossils of the giant hyena, *Pachycrocuta brevirrostris*. The traditional interpretation that humans occupied the main cavity has been challenged in recent years, and the alternative scenario is that the main occupants were the hyenas, so that the bones of ungulates, previously thought to be the hunting tro-phies of hominids, would instead have been dragged in by the hyenids, like the human bones (Boaz et al. 2000).

Lagoa Santa Caves

There are many other examples of sabertooth fossils found in cave de-posits, although not necessarily as part of fossil assemblages in dens. As we saw at the beginning of the book, large mammal fossils from Lagoa Santa caves in Brazil were thought by Lund (1842) to be the remains of the prey of *Smilodon*, which the latter would bring back to the cave to eat at leisure. But current views favor another scenario, in which the bones, including those of the sabertooth, came to the caves in different ways. For example, some animals died on the surface, and their bones were dragged to the cave floor by water currents. Other animals might have entered alive, seeking shade, water, or salt licks, and simply got lost and died inside (Cartelle 1994).

Carnivore Traps

Sinkholes, caves, and other types of cavities may act as natural traps for mammals, but in some cases the accidental victims of such traps may act as bait for other animals, attracting predators that enter the trap to scavenge but are unable to escape. In such cases, the sites are called car-nivore traps. But not all carnivore traps are cavities, and the most famous of them, the so-called tar pits (actually asphalt seeps) of Rancho la Brea, accumulated its fossils in a rather unusual way.

Rancho La Brea

As mentioned in chapter 1, Rancho La Brea is the site that yielded the first large collection of fossils of the Pleistocene sabertooth *Smilodon* in North America, in the early twentieth century, and fossils have continued to be found since then. The site has produced many thousands of *Smilodon* bones, by far the largest known sample of sabertooth fossils on the planet. The site is located in Los Angeles, California, and it encompasses over a hundred individual pits that are the result of excavations at the site since 1901. The sediments were accumulated in the late Pleistocene, between some 40,000 and 10,000 years ago, and were transported by alluvial fans originating in the nearby Santa Monica range. The oil that was pressed up from deep deposits then soaked the sands, creating a sticky, probably shallow, mass. Based on careful analysis of the fossils collected in controlled conditions over the last three decades from one of the pits (Pit 91), researchers have concluded that the skeletons of trapped mammals got buried rather quickly, but not before carnivores managed to take away many of the limb bones from the exposed side of the carcasses (Spencer et al. 2003). Some of those carnivores didn't get away with their prizes; instead, they themselves were trapped, becoming additional bait. In fact, it seems that the carnivore carcasses were scavenged almost as intensively as those of herbivores, which is striking because modern predators generally prefer not to eat carnivore flesh. As a result, Rancho la Brea has preserved an unusually high proportion of bones of carnivores, which make up more than 90 percent of the total vertebrate fossils. Secondary movement of the bones within the sediments due to limited flow has led to a thorough mixing of the bones in each pit, so that it is almost impossible to assemble a skeleton with the certainty that all the bones belong to a single individual (figure 2.2). Consequently, the body proportions of the mammals from La Brea, including the sabertooths, had to be inferred, with mean measurements calculated from dozens of specimens of each bone. It was only in 1986, during salvage excavations associated with the building of the G. C. Page Museum of La Brea Discoveries, when an associated skeleton ("associated" in this context means that the different parts of the skeleton are not mixed with those of other individuals and are found close enough together to make it clear that they belonged to the same animal) of *Smilodon* was found for the first time since the beginning of the excavations in the early twentieth century (Cox and Jefferson 1988).

The fossil fauna from Rancho la Brea is quite rich, and beside the sabertooths researchers have found there fossils of other large carnivores (including the American lion and the dire wolf) and a variety of herbivores (including camels, bison, antelope, deer, horses, giant ground sloths, mammoths, and mastodons), as well as a range of birds and smaller vertebrates. Together with pollen remains and other evidence, this fauna indicates a temperate climate, slightly cooler and more humid than today, and a mosaic of vegetation with abundance of pines, sagebrush and buckwheat.

Talara

Another carnivore trap resembling Rancho La Brea to some degree is the Talara Tar Seep in Peru. Like Rancho La Brea, Talara is of late Pleistocene age, and its dominant large predators are similar or comparable to

2.2. Schematic cutaway view of one of the asphalt seeps at Rancho la Brea (California) to show how the fossils probably accumulated. Top: a bison unintentionally steps on top of the seep. Center: the bison has been trapped in the asphalt, and carnivores gather around it to scavenge, several of them eventually getting trapped as well. Bottom: the bones of the trapped animals have been buried and mixed together by movement of the sediment.

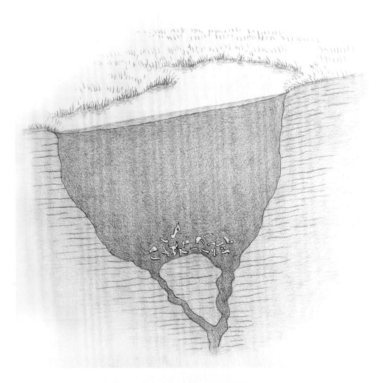

those at La Brea: the sabertooth *Smilodon*, a large jaguar *Panthera onca* (previously misidentified as an American lion, *Panthera atrox*), and the dire wolf (Lemon and Churcher 1961; Seymour 1983). However, it appears that the entrapment conditions were slightly different because the proportion of broken bones is higher in Talara, and many of those bones show signs of considerable weathering before burial. One explanation for these differences could be that the layers of asphalt-soaked sands were shallower in Talara than in Rancho la Brea, and the thrashing of large trapped animals caused more damage to the bones of previous victims. A high proportion of the trapped carnivores were juveniles, and many of the dire wolf specimens show skeletal pathologies, factors suggesting that lack of experience and a less than fit condition could increase an animal's chance of being trapped. An excellent sample of *Smilodon* fossils from Talara was taken to the Museum of Toronto, in Ontario, in the 1950s, but to this day the collection remains frustratingly undescribed.

El Breal de Orocual

The rich oil deposits in Venezuela have led to the formation of several "tar seeps," which only in recent years have been found to contain the remains of sabertooths. One of these sites, called El Breal de Orocual (in the northern Venezuelan state of Monagas), dates from the early to middle Pleistocene and thus differs from many other tar seep sites, which are from the late Pleistocene (Rincón et al. 2011). The site is especially important because it has produced the first clear record of the genus *Homotherium* in South America, not only proving that these animals actually entered the continent but also showing that their arrival was at a very early date.

Cave Traps

The most classic example of carnivore traps are caves into which herbivores fell accidentally through steep shafts, their carcasses attracting hungry carnivores that jumped in and were either killed in the fall or unable to escape.

Bolt's Farm

At Bolt's Farm, a site of Pliocene age (about 3 Ma) in South Africa's Sterkfontein Valley, the skeletons of three felid sabertooths of the genus *Dinofelis* were found together with the remains of about a dozen baboons. The cavity apparently acted as a trap for primates and carnivores alike, and it even contains coprolites, or fossil scats, of both types of animals, indicating that they survived for some time in the cavity (Cooke 1991). The absence of fossils of other large mammals suggest that only the cats and monkeys were agile enough to gain access to the cavity, but evidently some of them never managed to leave (figure 2.3). It doesn't seem that

2.3. A Pliocene scene within the natural trap in the cave site at Bolt's Farm (South Africa). The sabertooth *Dinofelis* confronts a baboon, which has also been trapped in the steep-walled cavity. Neither animal will be able to leave.

the carcasses of the first victims started a "chain reaction" of carnivore trappings, so this is not a carnivore trap in the full sense.

Incarcal

Another site that may have operated partly as a carnivore trap is Incarcal, in Girona, in northeastern Spain. This is a complex of karst cavities (nine cavities are currently known) in Pliocene lacustrine limestone filled with fossiliferous sediments. Similar cavities, or sinkholes, exist today in the vicinity of Lake Banyoles, not far from Incarcal, and they fill up when the water table is high. The holes may be encountered suddenly in the forest floor, and one has to be careful not to slip on the litter of dead leaves covering the inclined floor around the mouth of the hole, or a tricky fall would follow – and escape from a sinkhole, with its overhanging walls, is no easy task. The accumulation of fossils at Incarcal occurred at a time when the water column in the bottom of the cavities was low, and the sediments that entombed the remains were most likely to have come from outside the cavities during seasonal floods (Galobart 2003). The medium was partly anoxic due to its high sulphate content, creating an environment that was basic rather than acidic, and thus favorable to fossilization. Three of the Incarcal cavities have yielded the majority of fossils, including one of the best samples ever found of the sabertooth *Homotherium latidens*.

The Incarcal faunal association includes partial articulated skeletons of both the southern mammoth *Mammuthus meridionalis* and the hippopotamus *Hippopotamus major*. The presence of such skeletons indicates that the large ungulates were buried while ligaments still held their bones together, which – combined with the unusual abundance of carnivore remains (including *Homotherium* and the giant hyena *Pachycrocuta*) – suggest that the herbivores got caught in the bottom of the cavity, and carnivores gathered around the place to scavenge. One problem with the carnivore trap hypothesis is that the skeletons of the carnivores are not nearly complete, so it is necessary to identify what took so much of them out of the cavity. It may be that water currents acted on the carcasses. Another possibility is that the herbivores died near the cavities and their carcasses were dragged in by floodwaters, together with the remains of other animals, but this scenario hardly accounts for the abundance of carnivores, which in that case would have left the area undamaged instead of dying there along with their prey.

Batallones

Perhaps the most spectacular example of a true carnivore trap is the Cerro de los Batallones site complex in Madrid, in central Spain (figure 2.4). This is a series of cavities that formed during the Vallesian (late Miocene), at some distance from the borders of a large lake (Antón and Morales 2000; Morales et al. 2008). Even in severe droughts, the cavities held some standing water. Thus, they acted as traps for large ungulates such as rhinoceroses, giraffids, and mastodons, which would approach the cavities attracted by the moisture, enter them by choice or fall in, and then be unable to get out due to the shape and texture of the walls. In at least some of the cavities, the carcasses of the trapped herbivores acted as bait that attracted local carnivores, including the sabertooth felids *Machairodus aphanistus* and *Promegantereon ogygia*, conical-toothed cats of smaller size, bear-dogs, small hyenids, bears, and *Simocyon batalleri* (a leopard-sized relative of the red panda), as well as a variety of mustelids. At two of the sites, Batallones 1 and 3, the remains of carnivores make up over 90 percent of the mammalian fossils.

It is interesting to note that the majority of the fossils of *Promegantereon* from Batallones 1 belong to young adult animals. They were at a critical stage in their development, being pressed to leave their mothers' territories, and due to that pressure and their lack of experience, they were more likely to take risks than older animals. Such a bias is common in carnivore samples from trap sites, and it is mirrored in the high frequency of young animals caught in baited traps used for studies of modern leopards in Africa (Bailey 1993).

At another site, Batallones 2, the main find consisted of the skeletons of two mastodons, but drilling below where the mastodons were found indicates that several meters farther down is an accumulation of fossils dominated by carnivore remains. This suggests a pattern in the evolution of the cavities, which may have functioned as carnivore traps when they were deep and steep-walled enough. As they filled with sediment and became shallower, large mammals such as giraffids and mastodons

still were trapped, but the agile carnivores were not. The sediments that buried the mastodon remains from Batallones 2, or the giraffid skeleton from Batallones 4, must have flooded in rather quickly, not leaving much time for the carnivores to act on the carcasses; otherwise the herbivore skeletons would not be so complete and articulated.

Before the advent of paleontology, early findings of vertebrate fossils were interpreted as the remains of the victims of the biblical flood, which is why the term "antediluvian" ("before the Flood") was traditionally applied to any prehistoric creature. In geological history, floods have indeed been a major element in the preservation of fossils, and they must have killed millions of land animals. Not only exceptional, catastrophic floods helped to create fossils. In fact, the structure of the sediments in most fossil sites reflects regular cycles of drought and inundation. Many animals whose remains were buried by flood-transported sediments were not killed by the floods but probably died because of, or at least during, a previous drought.

The majority of land vertebrate remains that we find as fossils were originally accumulated in riverbed (fluvial), lakeside (lacustrine), or flood-plain deposits. The presence of adjacent highlands where erosion was taking place, with sediments flowing downhill, is required for fossils to accumulate. Of course, animals die every day in other environments such as forest floors, but the soil there is so acidic that any bones not destroyed by scavengers will ultimately disintegrate (figure 2.5). The sequence of events that leads to preservation of fossils in fluvial or lacustrine deposits can be summarized as follows: Animals die not far from the borders of a lake or river that is subject to considerable fluctuation, in most cases because of seasonal rainfall patterns. Most of the deaths occur during dry phases, when thirsty animals come to the drying river channels or water holes, and, finding no water, finally collapse and die. Others, in their weakened condition, are easy victims for predators. Later, water flows again, carrying fresh sediments and burying the remains of the animals under them. In other cases a flash flood would surprise animals, even entire herds gathered around drying riverbeds, drowning them and dragging the carcasses downstream.

This sequence contains variations—in the time the carcasses spend exposed to the elements before burial, the amount of energy in the currents, and the distance that the remains are dragged before being buried. Remains are best preserved when the time of exposure is short, the energy of the current is low, and the distance of transport is short. If carcasses remain exposed too long, they get weathered, dispersed, and ultimately destroyed by scavengers. If the current is too strong, the bones can be disarticulated and even broken. And if remains are transported over a long distance, they can be completely dispersed.

It is rare for the right variables for the best preservation to occur, and even then another series of lucky circumstances is needed. The sediments where the bones are buried have to escape various agents of destruction (essentially, metamorphism and erosion) until the present day, but then at least some erosion must occur to make the fossils visible and accessible for excavation. Thousands of perfectly preserved fossils surely lie underground in many parts of the planet, but if the sediments containing them aren't exposed by some erosive force, we will never come across them.

Flood-Plain, Fluvial, and Lacustrine Deposits

2.4. A scene in the natural trap of Batallones-1 (Spain), during the Vallesian (late Miocene, some nine Ma). Two sabertooths of the species *Machairodus aphanistus* snarl at each other over the carcass of a rhinoceros. The floor of the cavity is littered with the bones of animals that had been trapped previously.

2.5. A forest scene in the European Villafranchian (late Pliocene), with the felid sabertooth *Megantereon cultridens*. The body proportions of this animal indicate that it was a good climber and would have preferred forested environments. Although it had an enormous geographical range, relatively few remains of this sabertooth have been found – probably in part because forests are unfavorable environments for the preservation of fossils.

Against those odds, open-air fossil sites have yielded a huge amount of mammalian fossils, some of them in the form of complete, even articulated, skeletons. But most fossil finds are of fragmentary, dissociated bones that have to be painstakingly restored and put together again.

Some additional factors improve the chances of preservation, including volcanic activity. Many of the best sabertooth specimens were found in sediments containing a high proportion of volcanic ash. These cinders would be originally spread by the wind over enormous distances; they would then simply accumulate on the ground or be transported and rearranged by water currents, which lay them again in a stratified pattern. The chemical composition of the cinders makes the sediments less acidic and thus more suitable for bone preservation. In some cases the eruptions responsible for the production of the cinders actually killed the animals that later were entombed in the pyroclastic sediments.

Senèze

One place where the consequences of volcanic activity resulted in the exceptional preservation of sabertooth fossils is the early Pleistocene site of Senèze, in central France. Senèze is the caldera of an ancient volcano that exploded in the early Pliocene, leaving a maar (a large crater filled by a lake)—which, at some 500 meters across, was something like a miniature version of today's Ngorongoro Crater, in Tanzania. Later eruptions of nearby volcanoes spread layers of cinders that created the pyroclastic sediments where the fossils are entombed. The early excavations from the 1920s, which yielded the best skeletons known to date of the sabertooths *Homotherium latidens* and *Megantereon cultridens*, did not follow a careful methodology, and the original conditions of the fossil accumulation remained a mystery for decades. Over three-quarters of a century after its discovery, the site was re-excavated with more modern techniques, producing a detailed interpretation of how the animals died and became so beautifully preserved (Delson et al. 2006). It now seems that mudslides flowing down the inner walls of the caldera (possibly related to local faulting—meaning the development of fault lines) trapped the animals and dragged them to the lake borders, where their skeletons remained articulated and were preserved without being disturbed by scavengers.

Lothagam

Another open-air deposit that has yielded exceptional sabertooth fossils is Lothagam, a Miocene site located on the western shore of Lake Turkana, Kenya (Leakey and Harris 2003). One of the most spectacular finds is the nearly complete, articulated skeleton of a felid sabertooth that was classified in its own genus and species, *Lokotunjailurus emageritus*. The conditions in which the skeleton was deposited probably involved a large, meandering river subject to considerable seasonal fluctuations; and the sediments are largely volcanic. The fossils accumulated in a fine-grained, hard matrix that makes excavation laborious, but that contributes to an exquisite preservation. The subsequent faulting that occurred in the area lifted the whole Lothagam block, which is roughly ten kilometers long and sixteen wide, a full 200 meters above the surrounding plain, making the fossiliferous sediments accessible for excavation. Because of its fluviatile (a term referring to deposits formed by rivers) nature, the site has yielded many crocodile, turtle, and hippopotamus fossils, but it also has abundant remains of hipparionine horses, rhinoceroses, antelopes, proboscideans, and many other vertebrates.

Coffee Ranch

Roughly similar in age to Lothagam, Coffee Ranch is a classic open-air site known since 1930 It has yielded exceptional fossils of the North American felid sabertooth *Amphimachairodus coloradensis*, among a rich mammalian fauna that served to define the Hemphillian Land Mammal age of the American late Miocene (Evernden et al. 1964). The site was a lake basin, and the bodies of animals that died around the lake borders were eventually buried by flood-transported sediments. But the carcasses spent some time exposed, and the site provides a record of events that took place before their burial. A large slab of hardened mud preserved the footprints of several scavenging dogs of the genus *Borophagus*, which were moving around the location of an isolated ungulate rib. The tracks were then crossed over by the footprints of a sabertooth (one of the few examples of sabertooth fossil footprints), which apparently walked in a relaxed manner, oblivious to the presence of any other carnivores. Nearby are small piles of almost powdered bone, which seem to be the remains of the dogs' scats. Scavenging activity is also suggested by a pattern of damage to some of the bones (especially in the articulated sabertooth skeleton) which might correspond to the pecking of vultures. Greater damage to the bones was caused when they were trampled by large animals like rhinoceroses. Especially regrettable is the fact that one rhino apparently planted its heavy foot on top of the skull of the articulated sabertooth, hopelessly crushing its central part.

After the animal bones were entombed, successive layers of volcanic ash were washed by heavy rains into the lake basin, accumulating on top of the bone-bearing sediments and finally burying them under three meters of cinders, sealing the deposits and acting "like a cork in a bottle" (Dalquest 1969:3).

Sansan

Of earlier, middle-Miocene age is Sansan, in southern France. This is one of the most classic Tertiary sites, and it has been intensively studied since its discovery in 1834 (Ginsburg 1961a, 2000). New excavations and the analyses of old material with new methodologies are continually improving our knowledge of the Sansan biota (Peigné and Sen in press). The site was formed in a bend in a river or an oxbow lake surrounded by subtropical forest, which gave way to open woodland farther from the water. In the dry season, the flow would be minimal, but during seasonal floods, water would enter the lake or bend from the main river, dragging in all sorts of debris – including the bodies of mammals and other vertebrates. As the waters ebbed, the carcasses were stranded and finally entombed in the muddy sediments, which have preserved the remains of everything from birds to proboscideans. The animal whose genus name refers to the site is the sabertooth *Sansanosmilus*, an early kind of barbourofelid.

The Evolving World of the Sabertooths

Open-air flood-plain deposits have yielded the bulk of data that we use to reconstruct the history of our evolving biosphere, and there are hundreds of sites relevant for our understanding of the history of sabertooths in particular. In the previous paragraphs I have mentioned just a few of these sites, which provide clear examples of the features and processes typical of this kind of deposit. In the following sections, I offer a chronological review of the changing environments in which sabertooths of one kind or another evolved, punctuating the physical descriptions of those environments with mentions of particular fossil sites that shed light on concrete moments and scenarios in this story.

The Late Permian World

The first group of sabertoothed predators known to evolve were the gorgonopsian therapsids, which lived near the end of the Paleozoic, in a world very different from that of the present. In the late Permian, some 250 Ma, all the landmasses of the earth were part of a single supercontinent, Pangea. The climate was warm and continental, with contrasting dry and rainy seasons. Therapsids, which included both predatory and herbivorous forms, were the dominant land vertebrates, and they made up the first "modern" terrestrial community: as in modern ecosystems, vertebrates occupied both the herbivore and carnivore niches, whereas before the late Permian, the primary plant-eating animals were insects, and most terrestrial vertebrates were either insectivorous or carnivorous.

Gorgonopsian fossils have been found in several countries in sub-Saharan Africa and in European Russia. One of the best areas for finding them is the Karoo region of southern Africa (Catuneanu et al. 2005). The Karoo supergroup contains a series of stratigraphic units (that is, groups of sedimentary rocks that formed during the same time period), and it extends through much of present-day South Africa. Within the Karoo, the Permian and Triassic Beaufort group is the source of some of the best therapsid fossils, and it crops out in numerous sites in South Africa's Cape Province.

If we could fly over Permian landscapes of the Karoo, we would see river valleys with familiar features: sandbanks, flood plains, riverine forests, and swamps. But what looked familiar from the air would reveal itself as almost alien from the ground. The Permian landscapes would have been different not only to the eye but to the other senses as well. There was no such thing as bird song—the first bird was still over 100 million years in the future. The smell of wildflowers was lacking as well, with flowering plants even farther in the future than birds, so their buzzing cohort of pollinating insects was also absent although insects themselves were both present and prominent, some of them of gigantic size. And, of course, there was no grass. The flora of the southern parts of Pangea was dominated by forests of *Glossopteris*, early relatives of modern conifers. The understory of the forest and the flood plains were covered with smaller plants, including horsetails and ferns.

In the northern parts of the distribution of the gorgonopsians, including what is today Russia (Modesto and Rybczynski 2000; Ochev 2004), the vegetation was somewhat different, with the conspicuous absence of *Glossopteris*, its place taken by early conifers. Such vegetation provided food and shelter for a host of strange-looking herbivores, many of which belonged to the therapsid order Dicynodonta. Dicynodonts such as *Dicynodon* and *Lystrosaurus* were stocky animals, with long, barrel-like trunks and short legs. Their heads had a sharp, turtle-like beak and a single pair of tusk-like teeth. A more primitive group of therapsids, the Dinocephalians, included both carnivorous and herbivorous forms. A group of bigger, non-therapsid herbivorous reptiles were the paraeisasurs,

including *Paraeiasaurus* and the Russian genus *Scutosaurus*. The abundant vegetation around Permian watercourses not only provided food for herbivores, but it also offered an important resource for the big predators: cover. Sabertoothed "gorgons" such as *Rubidgea* from the Karoo or its Russian counterpart *Inostrancevia* might simply wait, hidden among the riparian vegetation, until the herbivores inevitably came to the shore to drink, or they could stalk unwary victims, taking advantage of the plant cover until they were near enough to dash in and attack (figure 2.6).

At the end of the Permian, the global environment changed for the worse, ultimately bringing about the disappearance of about 90 percent of all life-forms on earth, an extinction of greater magnitude than the one that would kill off the dinosaurs 180 Ma later. What sort of changes could trigger such a mass extinction? At least part of the cause seems to have been an incredibly intense episode of volcanic activity, lasting hundreds of thousands of years and leading to the accumulation of amazing amounts of volcanic rock in the area that today is Siberia, a volcanic mass known as the Siberian Traps. Such monstrous volcanic activity could cause worldwide acid rains, killing most of the planet's forests and unleashing chain reactions that ultimately led to the extinction of many land vertebrates. The massive death of forests is indicated by the presence in several parts of the world of end-Permian sediments with enormous concentrations of wood-eating fungi, suggesting huge amounts of dead trees. A recent study of riverine sediments around the time of the extinction in the South African Karoo has shown a change from a meandering river with a wet flood plain to a less meandering river with a dry flood plain, a change that took place simultaneously with the extinction of many species of vertebrates. This suggests an increased aridity and associated erosion of the soil. Other evidence indicates a continuous soaring of temperatures over the last few hundred thousand years of the Permian, which coincided with the volcanic activity in Siberia and was probably associated with the greenhouse effect of increased atmospheric carbon dioxide from the volcanic emissions. Nearly stagnant, oxygen-poor seas would also contribute to environmental deterioration (Ward et al. 2005). Among all that environmental deterioration and widespread extinction, one genus of dicynodonts, *Lystrosaurus*, managed to survive, and it actually thrived right after the extinction event, becoming the single most abundant land vertebrate. Its moderate size, burrowing habits, and ability to feed on extremely tough vegetation apparently allowed it the key to survive when almost all other terrestrial animals, including its feared predators, failed (figure 2.7).

The Non-Sabertooth Hiatus

After the end-Permian mass extinction, a full 200 million years passed before the earth saw another sabertoothed predator. During this cosmic length of time, Pangea broke apart into continents that drifted thousands of kilometers, approaching, but not quite reaching, their current positions. The dinosaurs appeared, rose to dominance, and were wiped off the planet, apparently by the collision of a huge asteroid with the earth at the end of the Cretaceous. True mammals appeared at the beginning of the age of dinosaurs, but they remained humble creatures in the shadow of the ruling reptiles until the end-Cretaceous mass extinction offered them a land of evolutionary opportunity. Birds arose from theropod

2.6. A scene in the late Permian, in the Karoo region in South Africa. Two coyote-sized gorgonopsians of the genus *Aelurognathus* lose their rightful prey, a dicynodont, to their larger relatives of the genus *Rubidgea*. Ferns and *Glossopteris* trees are seen in the foreground, and in the background is a semiarid valley whose meandering river is surrounded by marshes and riverine woods.

2.7. A scene in the earliest Triassic, just after the end-Permian extinction event. A dicynodont of the genus *Lystrosaurus* contemplates the desolated landscape from the entrance of its burrow. The sun-bleached skeletons of dicynodonts of other species can be seen in the dry hills, and a river with an arid flood plain flows in the valley bottom.

dinosaur ancestors, and by the end of the Mesozoic they had diversified into many of the modern groups. Flowering plants appeared and became widespread in the Cretaceous, changing the appearance of landscapes and fostering the evolution of diverse pollinating insects.

After the Cretaceous extinction, the Tertiary period, also known as "the Age of Mammals," began, and mammals indeed enjoyed an evolutionary explosion. During the Paleocene, the first period of the Tertiary, they spread over lands largely covered by tropical forests, with crocodiles, turtles, and huge snakes as the most obvious reminders of the previous "Age of Reptiles." The planet's climate was warm and humid, and as suitable for the life of reptiles as we can imagine, but the reign of the dinosaurs was gone forever, except for their one surviving branch – the birds – which diversified in no less spectacular a manner than the mammals.

Paleocene mammals evolved into forms of larger size than their Mesozoic, dinosaur-fearing ancestors, but the largest were still no bigger than a bear; the true giants of the mammalian world were still far in the future. Predatory mammals were even smaller than the herbivores, but other creatures seized the opportunity to feast on mammalian flesh: as if in a sort of revenge of the dinosaurs, some ground birds evolved during the Paleocene into the dyatrimas – fearsome creatures that were two meters tall and armed with massive beaks, and that had a taste for meat. They persisted into the next epoch, the Eocene, a time when mammals evolved into a fantastic variety of forms including herbivores of full rhinoceros size, and serious predators. But by the time the first mammalian saber-tooths appeared in the middle Eocene, the dyatrimas were declining, and they soon became extinct. The early Eocene saw the peak, or climax, of the worldwide tropical climate, with rainforests spreading as far north as Alaska and with Antarctica covered by dense woods. But with the advent of the middle Eocene, a cold and arid period caused important changes in world vegetation and induced the extinction of many kinds of browsing mammals.

The Middle Eocene in North America

The first group of mammalian sabertooths to evolve were the machaeroidine creodonts, distant relatives of modern carnivores that lived some 50 Ma, in the middle Eocene. They inhabited a world that would look much more familiar to us than the alien landscapes of the Permian, but that would not feel quite like home for the modern inhabitants of the region where the sabertooth fossils have been found, in the Rocky Mountains area in temperate North America. Although the continents were approaching their current positions, the Eocene world map was strikingly different from the present one, and many mountain ranges were yet to appear or acquire their current shape. The Rocky Mountains were then just a young range, and the basins between the mountains were at much lower altitudes than today. Thick, tropical forests covered river basins in what is today Oregon, Wyoming, Utah, and Colorado. The Bridger Basin has yielded some of the most spectacular and beautifully preserved fossils of Eocene mammals, including the best remains of the sabertoothed creodont *Machaeroides*.

The Bridger Basin contains a succession of faunas spanning some 4 million years. In the Eocene and Oligocene, the erosion of the young Rocky Mountains provided abundant sediments (Murphey et al. 2011). Early in this period, the basin was dominated by a shallow lake, which was gradually filled in with volcanic deposits, which in turn were later dissected by a fluviatile system. Widespread layers of cinders that spread as far as 300 kilometers from the volcanic source are evidence of violent volcanic episodes. The cinders traveled downstream in a huge wave that instantly changed the whole architecture of the valley. These volcanic episodes are also marked by mass deaths of freshwater turtles, whose carapaces are found by the hundreds in the Bridger Formation.

The Bridger Basin landscape in the Eocene must have been an awesome sight, especially from the viewpoint of one of the flying birds that were so abundant. A large, winding river occupied the valley bottom, and oxbow lagoons formed in places where the flow was slower. In some parts, the forest would reach the very border of the river, while in places where the current was slower, swampy vegetation developed (figure 2.8). The lowlands were densely forested with a great variety of tree species, including willow, walnut, birch, oak, maple, palm, and many other trees. Pine forests covered the flanks of the mountains, and higher still reared the cones of active volcanoes. Seen from the viewpoint of *Machaeroides*, the size of a domestic cat, the world of the Bridger Basin

2.8. A scene in the middle Eocene in the Bridger Basin, with two individuals of the dinocerate *Uintatherium* and the creodont *Sinopa* (in the foreground).

was one of towering trees and abundant mammals moving through the light and shade of the forest floor. Some of these could be targeted as potential prey, including the fox-sized members of the genus *Orohippus*, ancestors of the horse. Many of the vegetarian mammals were simply too large to be considered prey, including the vaguely rhino-like uintatheres, with their elephantine, short-limbed bodies and bizarrely tusked, six-horned heads.

Some of the predators were smallish creatures, including hyaenodontid creodonts, but others were larger. Among these were the stocky creodonts of the family Oxyaenidae, such as the formidable *Patriofelis*. It was as heavy as a tiger; had a huge head, shearing dentition for cutting meat, and short but frighteningly muscular limbs; and could easily displace *Machaeroides* from its rightful kills – in fact, it could almost swallow the smaller predator whole.

The other machairoidine genus, *Apataelurus*, lived in the Uintan, the next stage of the middle Eocene. The Uintan marks the beginning of the end of the "global greenhouse" that had characterized the first half of the Eocene, and this shift is reflected in the mammalian faunas. Tropical arboreal species become scarcer, and mammals adapted to subtropical and even temperate conditions and habitats appear. A marked transition in the evolutionary history of North American mammalian faunas also takes place during the Uintan, when about 30 percent of modern mammalian taxonomic families first appear in the fossil record, including the ancestors of modern carnivores, camelids, and some modern rodents (Murphey et al. 2011).

The Late Eocene and Oligocene World

The next group of mammalian sabertooths to evolve, the catlike nimravids, become abundant in the fossil record millions of years after the extinction of the last machaeroidines. Fragmentary fossils attributed to nimravids have been found in the middle Eocene of Asia, but it is only in the late Eocene of North America, in the epoch known as the Chadronian, that we find a well-documented record of this group. During the Eocene the climate changed considerably, and by the time the early nimravids appeared, temperatures had dropped, and the paradise of lush tropical forests that had covered much of North America and Eurasia during the first half of the period was giving way to drier, more open woodlands.

In North America, the best record of land mammals from the times of the nimravids continues to come from the Rocky Mountains area. After constant erosion, the mountains had become almost obliterated, and the whole area was now a plain some 700–1,000 meters above sea level, whose slow, meandering rivers continued to accumulate fine-grained sediments. An incomparable wealth of mammalian fossils has come to light in the White River deposits, which form extensive badlands over parts of North and South Dakota, Nebraska, Colorado, and Wyoming. The White River series includes the famous *Titanotherium* beds, which are of Chadronian (late Eocene) age, as well as the *Oreodon* beds, of Orellan or early Oligocene age, and the *Protoceras* beds, of Whitneyan or late Oligocene age (Hoganson et al. 1998).

By Chadronian times, when the sabertoothed nimravids *Hoplophoneus mentalis* and *Dinictis felina* are recorded, the forests that covered the Rocky Mountain basins had lost the lushness of the early Eocene, but still there were many tropical and subtropical trees. In the mammalian fauna, a turnover had taken place, and many archaic groups that were alive and well in the Uintan disappeared completely by the Chadronian. Among these were the six-horned uintatheres, replaced by a no less spectacular group of horned giants: the brontotheres, known from the late Eocene of both North America and Asia. Brontotheres belonged to the modern order Perissodactyla, but they looked like no living perissodactyl, with their massive bodies, high shoulders, and oddly horned heads (figure 2.9). There were many other large herbivores: several types of running rhinos galloped across the woodlands, looking like a cross between modern rhinos and horses; there were members of the extinct artiodactyl family Anthracotheridae, which were hippo-like but had long, narrow muzzles; and there were herds of early oreodonts, an extinct family of artiodactyls. Some oreodonts had vaguely sheep-like body proportions, but others were larger and more robust. Perhaps the strangest artiodactyls of the time were the entelodontids, including *Archaeotherium.* This animal had a pig-like

2.9. A scene in the Chadronian (late Eocene) in
Western North America, with
riverine vegetation. A nimravid
sabertooth of the species
Hoplophoneus mentalis is seen
in the foreground, while two
brontotheres of the species
Megacerops coloradensis
drink on the opposite side of
the river.

body, but with longer limbs, and its huge head displayed bony protuber-
ances on the cheeks and mandible (figure 2.10). Its dentition indicates an
omnivore diet, and it would readily scavenge if the opportunity arose. The
Chadron mammalian predators included medium-sized hyaenodonts
like *Hyaenodon*, ancestral bear-dogs such as *Daphoenus*, and smaller
hunters such as the earliest canids (members of the true dog lineage),
like *Hesperocyon*.

The leopard-sized sabertooth *Hoplophoneus mentalis* was the dominant catlike predator, and its only serious competitors were the larger
hyaenodonts. However, it is clear that the adults of the larger peryssodac-
tyls, including the brontotheres and many rhinos, would be essentially
free from predation.

The transition from the Chadron to the next epoch, the Orellan,
also marks the Eocene-Oligocene transition in North America, a time
marked by a strong climatic deterioration called the "Big Chill" (Prothero 1994:167). Such climatic changes had profound effects on the
vegetation and terrestrial environments, but nimravid sabertooths including *Hoplophoneus* and *Dinictis* continued to prosper for millions of
years, suggesting that they may have been more tolerant of dry, relatively
open environments than their robust proportions might suggest. By the
end of the Oligocene, the last American nimravids had disappeared, as

had almost all genera of large carnivorous mammals on the continent (Bryant 1996b).

2.10. A scene in the Oligocene in Villebramar, France, with two nimravids, *Eusmilus bidentatus*, retreating from the advance of two entelodons intent on scavenging the sabertooths' kill.

The European Late Eocene and Oligocene

In Europe the earliest nimravids are known from the early Oligocene, when the lynx-sized sabertooth *Eusmilus bidentatus* appeared in French and German sites. Europe had been a tropical archipelago for much of the Eocene. It was becoming closer to its modern shape as sea levels lowered and firm ground emerged, creating corridors of dry land that joined former islands and allowed land mammals to migrate. New bridges connected Europe with the larger landmass of Asia, and the ensuing invasion of terrestrial mammals had a devastating effect on the faunas that had evolved in relative isolation on the European islands. The changes in the composition of European mammalian faunas were so profound that the event has been termed "La Grande Coupure," or the "Great Cut" (Prothero 1994:189). The climate was becoming cooler and drier, and the lush tropical forests of the Eocene were giving way to more open woods and shrub lands. The new mammals that migrated from Asia had evolved in an environment that was strongly continental and thus already dry and seasonal, so they had an edge over the native European species. *Eusmilus* was one of these invaders, and with it came an impressive array of eastern mammals. These included primitive, running rhinoceroses, and the fearsome "giant pigs," or entelodons, which – in Europe as in North America – probably plagued the unimpressive predators of the time, appropriating their kills without ceremony (figure 2.10).

The Eurasian Miocene

The transition from Oligocene to Miocene marks the end of the first part of the Tertiary, what geologists know as the Paleogene, and the beginning of the second half, known as the Neogene. In the early Miocene, a new group of sabertoothed carnivorans enters the scene: the barbourofelids. Their origins probably lie in Africa, but they soon invaded Europe, which at the time enjoyed a benign climate and – after the cool, arid spell of the Oligocene – was once again largely covered in subtropical forests (Agustí and Antón 2002). Lynx-sized *Prosansanosmilus*, the first European barbourofelid, coexisted with a varied fauna including small and medium-sized ungulates, which were adequate potential prey. Among these were early ruminants such as the cervid *Procervulus*, and the first equid to arrive from North America, *Anchitherium*. These three-toed horses were browsers, and they flourished in the European forests of the early Miocene. Other newcomers were the first proboscideans, mastodons of the genus *Gomphotherium* that left Africa at the beginning of the Miocene. Rhinoceroses were abundant and varied, but just as unlikely to be on the menu of *Prosansanosmilus* as the mastodons. Among the predators, the amphicyonid bear-dogs were dominant, with the body size of a modern black bear; hemicyonine bears, which were fleeter of foot than their modern relatives, also attained large sizes. They were joined by another African immigrant, the giant creodont *Hyainailouros*, a relative of *Hyaenodon* but grown to full bear size, with an enormous head over half a meter in length. This animal was evidently an active predator, but because it could also deal well with bone, it was an adept scavenger. With so many large predators around, a conflict around a carcass in early Miocene Europe could be a violent one. Even if the carcass under dispute was the prey of

Prosansanosmilus, the best place for the smallish sabertooth to be during such conflicts would be as far away as possible, or at least high up the branches of the nearest tree.

The next barbourofelid genus to live in Europe, *Sansanosmilus*, is known from sites of middle to late Miocene age, a time when the world climate cooled further and the extension of the Antarctic ice cap increased. European forests grew thinner, and open woodlands with larger grassy patches became dominant features of the landscape. One of the best portraits of life in that period comes from the French site of Sansan, from which came the name of the sabertooth (see above). This site opens a veritable window to the middle Miocene world: in addition to a long list of mammalian species, it has also yielded a remarkably rich avifauna, including pheasants, owls, and various waterfowl. All these animals lived in a warm, seasonal, subtropical forest (Ginsburg 2000). *Sansanosmilus* itself was a leopard-sized predator and thus larger than *Prosansanosmilus*, but other predators around it had grown larger, too. *Amphicyon major*, a bear-dog typical of the middle Miocene, was as large as a brown bear, and hemicyonines like *Phlytocyon* were almost as large as a lion, fast and fearsome. A more direct competitor of *Sansanosmilus* was the true felid *Pseudaelurus quadridentatus*, a leopard-sized cat with very slight sabertooth adaptations. New species of herbivorous mammals such as the cervids *Euprox* and *Heteroprox* (figure 2.11) had also appeared, and they

2.11. A scene in the middle Miocene in Sansan, France, with the barbourofelid *Sansanosmilus* hunting the early deer *Heteroprox.*

would be on the menu of *Sansanosmilus*. The diversity of pig species also increased, with the genus *Listriodon*, armed with spectacularly curved tusks in the males, appearing at this time.

The Vallesian: Beginning of the Old World Late Miocene

The late Miocene saw a continuous retreat of forests and their replacement by open woodlands with grassy patches, resembling those seen today in some regions of India that are dominated by a monsoonal climate. Mammals that had already adapted to such opening habitats invaded Europe from the East, including the three-toed horses of the genus *Hipparion* (Agustí and Antón 2002; Bernor et al. 1997). Broadly similar in size to modern African zebras, they were adapted to both browsing and grazing. *Hipparion* became enormously widespread after its initial immigration from North America some 10 Ma, so that the Eurasian faunas of that age until the end of the Miocene are informally known as the "*Hipparion* faunas." Bovids (antelope) also became widespread, as did the giraffids and rhinoceroses of modern type, and this combination of elements gave these late Miocene faunas an "African" air quite different from modern Eurasian faunas, where the group of large herbivore species is largely made up of deer and wild cattle.

During the Vallesian, barbourofelids went extinct in Eurasia, and sabertooths of the "true cat" family Felidae rose to dominance. Among the first genera of felid sabertooths were *Machairodus* and *Promegantereon*, and the best samples of fossils of these animals come from the Spanish site of Cerro Batallones, as discussed above in this chapter (figure 2.12). But other sites in Europe complement the picture of Vallesian faunas and environments. Still in the Iberian Peninsula, the site of Can Llobateres in the Vallés Basin has yielded one of the richest faunas, with a huge variety of herbivores and with both felid (*Machairodus*) and barbourofelid (*Sansanosmilus* or *Albanosmilus*) sabertooths, documenting the coexistence of such potential competitors. The kind of broadleaf forests (or Laurisilvas) that covered eastern Spain at that time provided a rich mosaic of vegetation, supporting a wide variety of environments and prey.

Another wonderful example is Höwenegg, in Germany, one of the few sites where complete skeletons of large mammals have been found articulated and preserving traces of soft tissue outlines (Bernor et al. 1997). Unfortunately, sabertooth fossils from Höwenegg (again, *Sansanosmilus* and *Machairodus*) are scarce and fragmentary, but some of the sabertooths' prey animals, such as the horse *Hipparion* or the antelope *Miotragocerus*, are represented by amazingly complete skeletons, including a *Hipparion* mare with the remains of a fetus still within the contour of her belly.

The Turolian

The next epoch, the Turolian, saw an extension of grasslands in Eurasia, and as the vegetational cover decreased, so did the diversity of mammals. But if Turolian faunas were less diverse than Vallesian ones, they were no less spectacular. The dominant sabertooth of Turolian faunas of Eurasia was the felid *Amphimachairodus giganteus*, which was first found in one of the most famous European fossil sites, Pikermi (Solounias et al. 2010). This locality, a few kilometers northwest of Athens, has been excavated since the mid-nineteenth century. Fossils are found in accumulations that probably formed as water currents piled up carcasses in shallow ponds on an ancient flood plain. The main features of the accumulation were described by the French paleontologist A. Gaudry in 1862: "The most abundant remains are those of rhinoceroses, antelopes, and especially *Hipparion*. They are found in the greatest disorder; a rhinoceros skull may be hiding that of a monkey, and the monkey's limbs may be found beside the skull of a carnivore. It is rare that the bones of the same individual are found in connection" (Gaudry 1862:14). Even so, the bones are in good condition, indicating brief exposure and short transport—and some anatomical connections do appear. One of these included the bones of the forepaw of a sabertooth preserved down to the claws, including the enormous dewclaw typical of machairodontines.

A recent analysis of microwear on ungulate teeth from Pikermi reveals a diversity of diet adaptations, including browsers, mixed feeders, and grazers. This suggests that the site had a mosaic of vegetation dominated by open woodland, with patches of grass here and there (Solounias et al. 2010).

2.12. A scene in the open woodlands of the Vallesian (late Miocene) in Batallones, in central Spain. From left to right: the moschid *Micromeryx;* the amphicyonid beardog *Magericyon* (in the foreground); the felid sabertooth *Promegantereon* (up in the tree); the bear *Indarctos* (background); the hornless rhinoceros *Aceratherium;* the suid *Microstonyx* (background); the mastodon *Tetralophodon;* the three-toed horse *Hipparion* (or *Hippotherium*); the felid sabertooth *Machairodus;* an undetermined sivatherine giraffid (background); an undetermined boselaphine antelope (background); and the small hyenid *Protictitherium* (foreground). The holes in the ground are cracks in the flintstone, opening into the pseudo-karstic cavities that trapped the animals, mostly carnivores.

2.13. A panoramic view of the Turolian (late Miocene) environment at Venta del Moro, eastern Spain. From left to right: the hippopotamus *Hexaprotodon*, the mastodon *Anancus*, the bear *Agriotherium*, the felid sabertooth *Amphimachairodus* (on the carcass of the bovine *Parabos*), the felid sabertooth *Metailurus*, the hyaenid *Thalassictis*, the antelope *Tragoportax*, the camel *Paracamelus*, the three-toed horse *Hipparion*, and the castor *Dipoides*.

The best preserved skulls of *A. giganteus* known to us actually do not come from Pikermi but from the famous "Hipparion Beds" of China (Deng 2006). Such deposits occupy a large area in the northwest of Shanxi Province, and fossils are found there within a reddish matrix known as "red clays," remarkably similar to the sediments of Pikermi. Mammal bones are found in concentrations or "bone nests," which probably correspond to depressions in the flood plain of old fluvial systems where the bones would collect after being dragged by the floodwaters. The Chinese faunas are broadly similar to those from the Mediterranean, with many genera of mammals in common, although the species are different. Among the herbivores, grazing species predominate, and forest browsers are rare. Combined with other evidence, this indicates semi-arid steppe environments, rather than the open woodlands dominant in the Mediterranean area at that time (Deng 2006).

The end of the Turolian saw complex changes in the faunas of the Mediterranean area, with the arrival of immigrants from Africa (hippopotamuses), Asia (camels), and North America (canids) The migrations may have been a result of the general lowering of the sea level, which would have created new corridors of dry land on which mammals could move between continents. Life in the Turolian, some 6 Ma, is well portrayed in the fossil site of Venta del Moro, in Valencia, Spain (Morales 1984). A very rich fauna of large mammals has been recovered from this site, including the sabertooth *Amphimachairodus* and several other large carnivores such as the enormous bear *Agriotherium*, as well as many herbivores, such as antelopes, camels, horses, hippos, and mastodons of the genus *Anancus* (figure 2.13).

The end of the Turolian was marked by one of the most dramatic environmental changes in the Old World: the drying up of the Mediterranean Sea. As the connection with the Atlantic Ocean (around what is today the Gibraltar Straits) was interrupted, the waters of the

Mediterranean gradually evaporated until all that was left behind were endless, white extensions of salt from which the former islands reared like huge mountain ranges. No doubt it was fitting scenery for the end of a geologic period.

The North American Miocene

The upper Miocene in Europe had seen the decline and extinction of the barbourofelid sabertooths and the flourishing of the first machairodontine cats. But in the meantime the barbourofelids had invaded North America, where they would linger for millions of years, evolving into their most spectacular species.

In Miocene North America, the rain shadow created by the Rocky Mountains, combined with global climate changes, was causing a drastic reduction in forest cover, and the vegetation in the Great Plains became more and more open, ranging from savannah to steppe-like, with wood cover restricted to sheltered river valleys. Horses and herding rhinoceroses were the dominant herbivores, accompanied by oreodonts (which declined through the Miocene), camelids, and a variety of antilocaprids.

At the late Miocene site of Love Bone Bed, in Florida, abundant, if mostly unassociated, remains of the sabertooths *Barbourofelis* (a barbourofelid) and *Nimravides* (a felid) were found (Baskin 2005), together with a spectacularly varied fauna including both terrestrial and aquatic vertebrates (Webb et al. 1981). The fossils were collected upstream by floodwaters and finally gathered in a pronounced meander of a river that was part of a drainage system flowing into the Gulf of Mexico. This process mixed up the remains of creatures from different environments such as marshes, riverine woods, and grasslands, burying them in deposits of volcanic origin (figure 2.14).

2.14. A scene in the Clarendonian (late Miocene) in Love Bone Bed, in Florida. From left to right: the borophagine dog *Epicyon,* the short-legged rhino *Teleoceras,* the barbourofelid sabertooth *Barbourofelis,* the three-toed horse *Neohipparion,* the stilt-legged camel *Aepycamelus,* and the protoceratid *Syntethoceras.*

In younger deposits of Hemphillian age in Texas is the site of Coffee Ranch (see above), where abundant fossils of the sabertooth *Machairodus coloradensis,* including an articulated skeleton, have been recovered (Dalquest 1969, 1983). The environment around the paleolake (an ancient lake known to us only from the sediments that accumulated in it) was relatively dry and open, and subject to dramatic seasonal fluctuations. The fauna included other carnivores such as a pseudaelurine cat, a large bear, and a bone-eating dog, and herbivores including several species of camelids, antilocaprids, deer, horses, and rhinoceroses.

Africa in the Miocene

In the middle Miocene, when ancestral barbourofelid sabertooths are first recorded in Africa, the landscapes and vegetation of the continent differed from modern ones in important ways. The Great Rift, that giant valley crossing the eastern half of the continent from north to south, was still only an incipient feature. In part because this valley has had an impact on the distribution of winds and moisture, and in part because of differences in the global climate over time, the savannahs and grasslands

2.15. A woodland scene in the late Miocene in Lothagam, Kenya, with the early elephantid *Stegotetrabelodon* (background) and the felid sabertooth *Lokotunjailurus*.

2.16. A scene in the late Chapadmalalian (late Pliocene) in Argentina. From left to right: the glyptodontid *Paraglyptodon* (live animal in the background, carcass in the foreground), the ground sloth *Glossotheridium,* and the marsupial sabertooth *Thylacosmilus.*

were much more restricted than today, and a larger proportion of the continent was covered by woodlands (Turner and Antón 2004).

During the late Miocene, African vegetation became drier, with sparse woodlands and more extensive grasslands. At least three species of sabertooths, including one closely related to *Amphimachairodus giganteus,* have been found in the late Miocene site of Langebaanweg, in South Africa's Cape Province. The site corresponds to the flood plain and estuary of a river (corresponding to the current Berg) that flowed into the Atlantic Ocean. Carcasses of many animals that died in the flood plain were buried in sediments at the beginning of each rainy season, and just as at Coffee Ranch, the coprolites of carnivores, with their high bone content, were preserved near the bones of the ungulates, testifying to the carnivores' scavenging. In one dramatic example, the breaking down of bone fragments in a hyena's digestive tract was interrupted by the animal's death, so the etched, semi-digested bones of the prey were found associated with those of the hyena itself.

As discussed above, the fossil site of Lothagam in Kenya has yielded a rich mammalian fauna of late Miocene age, broadly contemporary with Langebaanweg (Leakey and Harris 2003). The depositional environment at Lothagam is that of a large meandering river, and the fauna recovered is enormously rich. In fact, several depositional episodes are recorded at the site, ranging from Miocene to Pleistocene in age, but the richest deposits (which yielded the sabertooth skeleton) correspond to the Nawata formation, of late Miocene age. At that time, the river's course was permanent but subject to strong seasonal fluctuations, with a dry season of about four months. It was surrounded by lush gallery woods (a

term referring to forests growing along the borders of rivers) that, farther from the water, gave way to wooded savannah and semideciduous thorn trees. The sabertooth from Lothagam, called *Lokotunjailurus emageritus*, would have had a wide choice of prey among the many species of antelope, giraffids, pigs, hipparionine horses, and it might even have hunted the hippos, rhinos, and young proboscideans (figure 2.15).

2.17. A scene in the early Pleistocene in the Lake Turkana area of Kenya. From left to right: the elephant *Elephas*, the antelope *Kobus,* the felid sabertooth *Homotherium,* and the horse *Equus*.

South America in the Pliocene

For much of the Tertiary, South America was an isolated continent, and its mammal fauna evolved into a multitude of autochthonous forms completely different from the groups that were developing in the interconnected landmasses of Africa, Eurasia, and North America. The roles played elsewhere by artiodactyls and perissodactyls were taken in South America by strange native ungulates such as the camel-like litopterns or the diverse notoungulates. Instead of large placental carnivores, South America harbored the Borhyaenoidea, a superfamily of predaceous marsupials that ranged from vaguely wolf-like forms to the sabertoothed thylacosmilids, but there were also swift terrestrial crocodiles of the family sebecosuchidae and swifter, predatory ground birds of the family Phorusrhacidae, which were up to two meters tall and armed with fearsome beaks. Aptly named "terror birds," these dinosaur-like creatures are thought to have out-competed the wolf-like carnivorous marsupials in the open environments.

During the Miocene and Pliocene, when the known history of marsupial sabertooths took place, the evolution of terrestrial environments

2.18. A scene in the early Pliocene at Aramis, in Ethiopia. From left to right: the antelope *Tragelaphus,* the suid *Nyan-zachoerus,* the deinotherid *Deinotherium,* the hominid *Ardipithecus,* and the felid sabertooth *Dinofelis.*

in South America was dramatically affected by the emergence of the Andes. Extending from north to south all along the western edge of the continent, the Andes created a huge rain shadow that became more severe as the mountains grew (Pascual et al. 1996). Thus, the forests that covered much of the continent at the beginning of the Tertiary began to thin, giving way first to open woodlands, then to savannahs crisscrossed by gallery woods along watercourses, and finally to pampas-like grasslands. The Chapadmalalian age of the Pliocene, which saw the flourishing of the marsupial sabertooth *Thylacosmilus,* also coincided with a drop in temperature of the world climate; together with an intensified pull of the Andean orogenesis (a term referring to the geological process leading to the rising of a mountain chain), this made the Pampean environments more arid. The mammalian herbivores in South America were largely grazers, adapted to cope with the tough vegetation; they included litop-terns and toxodontids. The edentates included grazing glyptodons and browsing ground sloths (figure 2.16).

Africa in the Pliocene and Pleistocene

In northern Kenya, both the eastern and western borders of Lake Turkana have yielded numerous localities of Pliocene and Pleistocene age, most fa-mous for their abundance of fossil hominids but also containing a very rich mammalian fauna (Turner and Antón 2004). These are typical flood-plain sites, where fossiliferous sediments accumulated for more than a million years while the basin was alternately occupied by a lake or crossed by a river. Although the Turkana sites have yielded a few articulated skeletons, most of the remains are isolated and fragmentary, and the sabertooths are no exception. Even so, skulls of *Homotherium* and *Megantereon* and partial skeletons of *Dinofelis* have been found (figure 2.17).

The Turkana Lake is fed by the Omo River, which flows southward from Ethiopia, and the river's borders are as richly fossiliferous as the lake's. The Omo fossil sites yield faunas similar to those of the Turkana, with fossils of hominids and many mammals, including the typical Pliocene sabertooths; they have also provided a rich plant fossil record, including trunks of many tree species that grew along the ancient river borders (Dechamps and Maes 1985).

The area known as the Afar Triangle in northern Ethiopia has yielded many sites of Pliocene and Pleistocene age, including Aramis, in the middle course of the Awash River (White et al. 2009). This site is famous for the discovery of the early hominid *Ardipithecus ramidus*, but it has also yielded a rich mammalian fauna that includes remains of the sabertooth *Dinofelis* (figure 2.18). Also in the Afar triangle is another hominid site, Hadar, where the famous skeleton of *Australopithecus afarensis*, known as Lucy, was found, and where sabertooths were present as well. Hadar was part of a basin in which a meandering river produced a shallow lake and deltaic marshes where an amazing variety of wildlife thrived.

Eurasia in the Pliocene and Pleistocene

The beginning of the Pliocene in Europe coincided with a rise in temperatures and sea levels, one result of which was the refilling of the Mediterranean. For years a monstrous waterfall must have roared in what is today the Straits of Gibraltar, as the waters of the Atlantic Ocean flowed into and gradually filled the dry sea. Forest vegetation spread once more, creating an ideal habitat for the jaguar-like sabertooths of the genus *Dinofelis*. In southern France, the Roussillon basin has been excavated since the nineteenth century and has yielded the remains of many Pliocene mammals, including straight-tusked mastodons of the genus *Anancus*, hipparion horses, and bovines of the genus *Leptobos*. The presence of giant tortoises (*Geochelone*) in southern France is a good indication of the warm climate that pervaded Europe in the early Pliocene (Agustí and Antón 2002).

At the late Pliocene (Villafranchian) site of St. Vallier, also in France, fossils of the sabertooths *Megantereon* and *Homotherium* have been recovered, together with thousands of other mammalian remains. The bones are found in lens-shaped concentrations in hardened loess (wind-blown sediments originating in the moraine areas in front of retreating glaciers), and recent taphonomic studies have shown that the accumulations were most likely the result of water currents transporting the carcasses until they were caught on the banks (Valli 2004). They could not have been transported for long distances, because associations of bones belonging to the same individual are frequently found, although these are rarely articulated. The carcasses were exposed long enough for scavengers to act on them before burial, as abundant marks and traces of carnivore activity are evident on many of the bones. Also abundant are the marks of porcupine teeth, a fact related to porcupines'

2.19. An early Pleistocene scene at Dmanisi, Georgia. From left to right: the felid sabertooth *Megantereon,* the hominid *Homo,* the mammoth *Mammuthus,* and the horse *Equus.*

habit of gnawing on bones to add to their diet the calcium necessary for the growth of their fearsome quills.

Of slightly younger age is the Plio-Pleistocene site of Senèze, in central France, where the best skeletons known to date of the sabertooths *Homotherium latidens* and *Megantereon cultridens* were found (as discussed above in this chapter). The rich fauna of this site include other carnivores such as early wolves; hyaenids; the giant cheetah *Acinonyx pardinensis,* tall as a lion; and many species of deer, antelope, horses, rhinoceros, and proboscideans. Combined with the pollen record, this faunal association points to a mixed environment of woodlands and prairies in a temperate climate, not very different from that of the present day. Even so, there were also terrestrial monkeys related to the present-day langurs of India, an element we normally associate with warmer climates (Delson et al. 2006).

Of similar age to Senèze is the Georgian site of Dmanisi, where fossils of *Megantereon* and *Homotherium* have been found – together with those of hominids and many other animals – in volcaniclastic sediments that accumulated between the course of an ancient river and the shores of a lake (figure 2.19). The interpretation of this site is complex: it has features of typical flood-plain sites, but there are also secondary cavities in the original sediment, which got filled with new sediment, itself highly fossiliferous. The bones appear to have been exposed for only a very short time, with water transport being minimal. Many bones bear clear traces of the activity of carnivores, and hominid tools also abound, so the attraction of meat played at least some part in the accumulation of the fossils. Since the site area was a sort of peninsula, given that it was surrounded by water

on three sides, it would have been a suitable place for predators, such as sabertooths, to ambush their prey. A variety of scavengers, including hominids, would have come to the place to share the meat, and sometimes they would become the victims of other predators during conflicts around the food. The remains of the predators' feasts and of the conflicts' victims would lie in the area until floodwaters dragged and accumulated them in the near-shore cavities (Gabunia et al. 2000).

2.20. A late Pleistocene scene in Alaska, with the felid sabertooth *Homotherium*. It is feasible that among the arctic populations of this sabertooth, white pelage was selected as an advantage in approaching prey undetected during the winter months.

The North American Pliocene and Pleistocene

After the end of the Miocene, *Amphimachairodus* disappeared from North America, but the typical Pliocene genera *Megantereon* and *Homotherium* arrived soon from Eurasia, and are known from several early Pliocene or Blancan faunas in North America. During the Pleistocene, *Megantereon* evolved locally into the genus *Smilodon*, while the place of the Blancan *Homotherium ischirus* is taken by *Homotherium serum* and

2.21. A late Pleistocene scene at Rancho la Brea, in Los Angeles. From left to right: the horse *Equus,* the bison, *Bison,* the felid sabertooth *Smilodon,* the mammoth *Mammuthus,* and the ground sloth *Nothrotheriops.*

the strange, robust homotherin *Xenosmilus hodsonae.* The success of these felid sabertooths is associated with the variety of habitats and large herbivores present in North America at the time, ranging from subtropical forests and savannah in Florida (home to the robust *Xenosmilus*) to the mammoth steppes of Alaska, where the gracile *Homotherium serum* shared the treeless landscape with lions and wolves.

Beyond the Arctic Circle, the permafrost of Alaska and Canada is rich in fossil sites where the bones and even mummified remains of ice age mammals are abundantly preserved. Many of these sites have been discovered as gold miners dismantled the frozen ground with pressure hoses and accidentally found the fossils. An area especially rich in fossils is the Old Crow River Basin in the Yukon (Kurtén and Anderson 1980). There, fossils of the sabertooth *Homotherium* have been discovered together with an amazingly varied fauna, including woolly mammoths, moose-stags, musk oxen, lions, short-faced bears, and many others (figure 2.20). The basin, traversed today by the meandering Old Crow River, was at several stages during the Pleistocene occupied by a large lake, as the rivers flowing eastward found their course blocked by glacier ice. Sediments that originally accumulated around that lake were later excavated by the river when the lake drained and erosion eventually revealed the fossils along the banks. Some of the fossil outcroppings have probably been eroding for thousands of years.

A rich portrait of the environments and fauna that surrounded the sabertooth *Smilodon* during the late Pleistocene in the more southern, temperate latitudes of North America is provided by Rancho la Brea, as discussed above. Horses, bison, camels, proboscideans, and giant ground

sloths grazed and browsed in the mosaic of woods, shrub land, and grass-land that covered southwestern North America during the last ice age (figure 2.21).

South America after the Interchange

During the late Pliocene and Pleistocene, the large mammal faunas of South America reached a balance between native and immigrant species, producing a spectacular and unique combination. Groups of northern origin such as mastodons, horses, camelids, and deer shared the prairies and woodlands of the once isolated continent with natives such as the toxodons, litopterns, giant ground sloths, and glyptodons (Pascual et al. 1996). With the last large predatory marsupials having been extinct since the Pliocene, the group of large predator species was made up of immigrant, placental carnivorans–including several species of large, hypercarnivorous canids; the giant short-faced bears of the genus *Arctodus*; and the large cats, both feline and machairodon-tine. The dirk-tooth *Smilodon* was enormously successful and appears to have evolved locally into the giant species *S. populator*, the largest sabertoothed carnivore ever. Scimitar-tooth cats of the genus *Homoth-erium* also entered South America, but their scarcity in the fossil record suggests that they never became very abundant in the continent (figure 2.22). This impressive assortment of large mammals persisted through the successive climatic oscillations of the Pleistocene, probably adjusting to the latitudinal changes in vegetation thanks to regional migrations. There was no major loss of species until the catastrophic, wholesale extinction at the end of the Pleistocene, shortly after modern humans became widespread in the continent.

2.22. A Lujanian (late Pleisto-cene) scene in Argentina. From left to right: the mastodon *Stegomastodon* (background), the glyptodontid *Glyptodon* (middle ground), the felid sabertooth *Smilodon,* the ground sloth *Megatherium,* the horse *Hippidion,* and the litoptern *Macrauchenia.*

As we have seen, sabertooths shared their habitats with a host of other predators, and just like in modern ecosystems, competition would have been a natural consequence of their coexistence. Each predator, like any other animal species, occupies its own ecological niche. A species' niche is its role in the ecosystem, its relations with food, predators, and other factors—all of which define its way of "doing business" in nature. In theory, each niche can be occupied by no more than one species at the same place and time. If two species with similar adaptations have to coexist, they will be thrown into direct competition with each other, and the slightest unbalance will serve to decide which species remains. As a long-term result of this phenomenon, called competitive exclusion, species are thought to evolve slightly different adaptations, from simple differences in body size to more complex morphological and behavioral changes that allow them to live side by side with other species, sharing the available resources. In the case of modern mammalian predators, it is very unlikely, if not impossible to find two similarly adapted carnivore species of the same body size in the same habitat.

In modern African savannahs, three species of big cats (lions, leopards, and cheetahs) share the same resource: ungulate meat. But differences in body size, anatomy, and habits allow them to avoid direct competition, at least in part. Leopards are smaller than lions and tend to take smaller prey, but they are also better climbers and can retreat to the safety of the high branches of trees (taking their food with them) when directly threatened by their larger cousins. Cheetahs are of similar or smaller body size than leopards, but they are more lightly built and better adapted for speed, which allows them to take the swiftest prey (usually gazelles) with greater efficiency than the other cats. But their more delicate physique makes them vulnerable to prey stealing and direct attack from the other, stronger carnivores. To avoid the latter, cheetahs tend to occupy more open sections of the habitat, and they usually hunt in the central hours of day, when lions and leopards are resting in the shade. Thus, although the cheetahs suffer frequent food stealing and aggression from their fellow predators, they are able to survive.

Although we often envision the African savannah as a grassy plain extending as far as the eye can see, the fact that the three big cats mentioned above can coexist there points to an important factor in this coexistence: the mosaic nature of the vegetation in the region. The open expanses with which we are so familiar are actually part of a patchwork of grasslands, woodlands, and gallery forests. In fact, the coexistence of more than one species of large felids normally occurs in habitats with some tree cover, as is the case here. The tree cover allows the smaller species to take refuge from the attack of the larger one and so avoid being killed or having its prey stolen (Morse 1974). This is the case not only for lions and leopards in Africa (Bailey 1993), but also for tigers and leopards in Asia (Seidensticker 1976). In the fossil record, whenever we find that a lion-sized big cat and a leopard-sized one appear together in a fossil site (and if the second is similar to the leopard in being a good climber) it can be taken as an indication that the environment around the site had enough vegetational cover for the small species to avoid the larger one. That is the case for sites like Cerro Batallones, with *Promegantereon and Machairodus*; Pikermi, with *Paramachaerodus* and *Amphimachairodus*; and for several Plio-Pleistocene sites around the world where *Megantereon* and *Homotherium* are found together. Of course the cheetah, with its greyhound-like build, is hardly a climbing cat, and it has a different solution for coexistence, one which allows it to exploit the other extreme of the vegetational spectrum. The very visibility that makes the cheetah vulnerable to prey-stealing from larger predators in the grasslands also allows it to detect approaching competitors and flee from the kill site in time to avoid being killed or mauled. But even the cheetah has a crucial need for cover when it has small cubs, whose survival largely depends on remaining hidden from view while their mother goes hunting.

Bearing in mind the tightly woven relationships that exist between the big cats of Africa and that allow them to share their resources, we might wonder what would happen if we introduced the American puma into the African savannah. Theory predicts that, given its similar body weight, habits, and adaptations, it would compete directly with the leopard. In such circumstances, one or both cats would have to evolve quickly enough into sufficiently different species – otherwise, one of them would be ousted from the region.

The same thing happens with other big predators. In Africa, wild dogs (*Lycaon pictus*) and spotted hyenas (*Crocuta crocuta*) compete for similar prey, and both are pursuit predators hunting in groups. But they are different enough to share their environment, even if some competition does occur. Hyenas, with their bone-crushing dentition, are better adapted to scavenging; the faster, lighter wild dogs are more efficient hunters; and thus the two species can divide local resources. But the hyenas take advantage of their greater bulk and large clans to bully wild dogs away from their rightful kills, so in places where visibility is high, wild dogs have a hard time trying to avoid the hyenas. The outcome of a direct conflict is not necessarily the immediate mauling of a dog, as would be the case with the solitary cheetah, because the wild dogs live in groups. But with a large enough group of hyenas, the loss of the prey is inevitable – which is almost as bad in the long term. One solution for the dogs is to take advantage of the cover in more wooded sections of the habitat, not so much to get closer to prey as to be able to eat from their kills for a longer time before being detected by their competitors (Creel and Creel 2002).

There are complex relationships among all the species of large carnivores sharing an environment, and these relationships help guide the evolution of each species (which, of course, is also determined by the species' phylogenetic constraints). In each ecosystem, the group of species that share the same resource and that have resolved the potential problems of

Sabertooths and Predator Guilds

competition form a more or less stable community, which in ecological terminology is known as a guild. The large predator guild is just one example; another is the guild of large ungulates, which shares the ecosystem with the predator guild and which in turn interacts with the latter over geologic time, in a phenomenon called co-evolution.

A review of the evolution of the large predator guild over the last 30 million years or so shows considerable stability in the diversity of trophic or alimentary adaptations. Large, predatory mammals have evolved adaptations for somewhat different diets, so while some of them eat almost exclusively meat (the hypercarnivores, such as the big cats), others (a group we may call "meat-bone" specialists, such as the hyenas) have more robust dentitions that allow them to crush and eat bones, thus including more carrion in their diet. A third group, called "meat-non-vertebrate" (including most dog species), have a more all-purpose dentition that allows them to eat a variety of food items, from invertebrates to vegetable matter (Van Valkenburgh 2007). At least since Oligocene times, it seems that these three major groups have managed to exploit their resources in a stable arrangement, but the fact is that individual species rarely last for more than a few million years. The balance of the guild is maintained because new species have filled in the niches vacated by the ones that vanished, but there are several ways in which this relay takes place. In some cases, a species invades an ecosystem and displaces an earlier species that filled the same niche but that, for some reason, was at a relative disadvantage. In other cases, the resident species goes extinct because of environmental pressures, and by the time a new species arrives, it finds the niche conveniently vacant; thus, no conflict takes place. In still other cases, a new species arrives in an ecosystem where a very similar niche is occupied, but the conditions are so favorable that the resident and the newcomer can coexist and share the resources. In such cases, unusually rich carnivore guilds appear, and can last as long as the balance of resources does.

There are examples of such rich guilds in the fossil record (figure 2.23). In recent times the wolf has been the dominant large carnivore in Europe, although lions were historically recorded in the southern parts of the continent. But during the late Pleistocene, both wolves and lions shared their prey with other predators now extinct in the area, like the spotted hyena, leopard, and cuon, and with others extinct everywhere, like the scimitar-tooth *Homotherium*. The latter is currently known from a single fossil dating from the last European ice age, about 28,000 years ago. However, before some 400,000 years ago it was quite widespread, and one can't help thinking that there must have been some drastic changes in the availability of resources to shrink the European large predator guild to its modern size (Antón et al. 2005). A similar situation occurred in Africa during the late Pliocene (Turner and Antón 2004), when the modern arrangement of carnivores—including lion, leopard, and striped and spotted hyenas—coexisted for some time with three different kinds of sabertooths (*Dinofelis*, *Homotherium*, and *Megantereon*) as well as two additional types of large hyena (*Chasmaporthetes* and *Pachycrocuta*).

But the fossil record also provides examples of the opposite situation. In North America, the sabertooth niche was vacant from the extinction of the last sabertoothed nimravids of the genus *Nimravus*, some 24 Ma, until the arrival of the first barbourofelids from Asia some 12 Ma, a long time even by geological standards. A similar situation occurred in Europe between the extinction of *Eusmilus* some 30 Ma and the immigration of barbourofelids around 19 Ma (Bryant 1996b). In South America, the extinction of marsupial sabertooths has been attributed to competition with the invading *Smilodon* from North America, but the last fossil record of the former (in the Chapadmalalian) is actually earlier than the first recorded appearance of the latter (Ensenadan, and some questionable Uquian finds), so it is likely that the thylacosmilids were already extinct by the time the felid sabertooths arrived. Whether or not they included sabertooths, some past ecosystems sustained remarkably small guilds of large carnivores, and in South America in particular the record of large marsupial predators is not especially impressive, leading to suggestions

2.23. Representative species of the early Pleistocene (top) and late Pleistocene (bottom) large carnivore guilds of East Africa. Note the greater species richness in the earlier guild.

Top, from left to right: ancestral lion (*Panthera sp.*), ancestral cheetah (*Acinonyx sp.*), leopard (*Panthera pardus*), spotted hyena (*Crocuta sp*), sabertooth cat (*Dinofelis sp.*), striped hyena (*Hyaena hyaena*), dirk-toothed cat (*Megantereon whitei*), and scimitar-toothed cat (*Homotherium hadarensis*).

Bottom, from left to right: cheetah (*Acinonyx jubatus*), leopard, lion (*Panthera leo*), spotted hyena (*Crocuta crocuta*), striped hyena, black-backed jackal (*Canis mesomelas*), and African wild dog (*Lycaon pictus*).

that the giant phorusrachid birds largely filled the ecological niches of large terrestrial predators. In view of the crucial role of large carnivores in present ecosystems, their paucity in some fossil communities is intriguing. Nonetheless, we must remember that lack of predator diversity does not imply lack of predation, and a few or even a single very efficient species of large predator can do a lot to keep herbivore populations at check–as the wolf has been doing in Europe and North America for several thousand years.

Sabertooth Ecomorphs

As we have seen, each kind of sabertooth displays its own set of peculiarities, but it is possible to see some patterns in their differences and similarities. In the mid-twentieth century, the Finnish paleontologist B. Kurtén grouped felid sabertooths into two broad categories, which he named "dirk-tooths" and "scimitar-tooths" (Kurtén 1968). As the names imply, these categories primarily reflect differences in the shape of the upper canines between the two machairodontine tribes, smilodontini and homotherini. As we shall see in the next chapters, the dirk-tooths (smilodontini) had very long sabers, with only moderate lateral compression and with very fine serrations, or none at all, on their margins. Scimitar-tooths (homotherini) possessed shorter and broad, but very flattened, sabers with coarsely serrated margins (figure 2.24).

But there is more than fang shape to these categories. As redefined years later by the American paleontologist L. Martin, the dirk-tooths and scimitar-tooths became full-fledged ecomorphs–a term that defines a set of morphological features that reflected a particular ecological niche. These ecomorphs are taxon-free definitions, meaning that any sabertooth meeting the right morphological requirements will qualify as a dirk-tooth or scimitar-tooth, no matter if it is a felid, a creodont, or a marsupial. According to this definition, dirk-tooths are characterized not only by their namesake upper canines, but also by having robust post-cranial skeletons, with short, bear-like limbs and plantigrade or semi-plantigrade feet. Scimitar-tooths had more gracile skeletons, with elongated limbs and fully digitigrade feet (Martin 1989).

According to this theory, such differences in morphology reflect two different ecological and behavioral solutions to the problem of being a sabertooth: the extremely muscular build and extra-long sabers of dirk-tooths corresponded to a solitary hunting style, with patient ambushing and short dashes aimed toward large, slow prey with thick skins. Such a hunting method would require the strong physique of a wrestler, and the long canines were adequate to pierce the skin of pachyderms and other big prey. Scimitar-tooths, thanks to their greater speed, would actively pursue fleeter prey, and in some cases they would act as a group. Accordingly, dirk-tooths would have a vital need of vegetational cover to ambush their prey, while scimitar-tooths would do well in relatively open country. During much of the Tertiary, the different predatory styles of dirk-tooths, scimitar-tooths, and conical-toothed cats would have allowed a neat partitioning of resources.

So the theory goes, and it may well reflect part of the truth, but reality is likely to have been more complex. On the one hand, most of the known mammalian sabertooths seem to fit better with the dirk-tooth model, while the scimitar-tooth combination of long limbs and broad, coarsely serrated sabers is clearly observed only in members of the Neogene felid tribe Homotherini. One purported non-homotherin scimitar-tooth is the Oligocene nimravid *Dinictis*, which indeed had short sabers and light, gracile limb bones, especially when compared with the contemporary, heavyset dirk-tooths of the nimravid genus *Hoplophoneus*. But the sabers of *Dinictis* were not more coarsely serrated or more flattened than those of *Hoplophoneus*, and although the former had comparatively longer forearms and shins than the latter, its metapodials (feet bones) were not markedly longer and hardly suggest reaching especially high speeds or sustained running in open habitats. Therefore, calling *Dinictis* a scimitar-tooth seems inappropriate (Scott and Jepsen 1936).

On the other hand, the Pleistocene homotherin *Xenosmilus* had unmistakable scimitar-tooth upper canines, but rather unexpectedly it was found to have the robust, short-limbed skeleton that we have come to associate with dirk-tooths like *Smilodon*. In fact the combination of adaptations in this animal was unique enough to lead the authors of its description to propose a new ecomorph for it, which they named "cookie-cutter cats" (Martin et al. 2011). Furthermore, if we look at some of the smaller dirk-tooth species—like the creodont *Machaeroides eothen*, the size of a domestic cat; or the lynx-sized nimravid *Eusmilus bidentatus*—it is hard to believe that they preyed on significantly more thick-skinned prey than their non-sabertoothed counterparts.

So it seems difficult to recognize the existence of clear-cut dirk-tooth and scimitar-tooth ecomorphs through the Tertiary history of mammalian sabertooths. Rather, it seems that the initial selection of sabertooth features took place among relatively powerful carnivores well able to hunt prey individually, and that the evolution of longer (higher-crowned) sabers was usually coupled with increasingly robust bodies fit for the immobilization of prey (Salesa et al. 2005; Meachen-Samuels 2012). In the particular case of the homotherin tribe of felids, the apparition of relatively short, flat sabers in an early, lion-sized form was coupled with the selection of an ever more gracile skeleton, related to preferences for an open habitat and probably group living, a lifestyle that involved partial loss of the individual's ability to immobilize prey and thus may have prevented the selection of especially long sabers. Then one genus in this group (*Xenosmilus*) probably shifted its habitat preference in favor of more closed environments, and a more robust physique was selected, giving rise to the apparently odd combination of characters of *Xenosmilus*. Although in the original ecomorph theory, sociality was an exclusive trait of scimitar-tooths, some large dirk-tooths that lived in open environments,

2.24. Comparison of the right-side upper canines of a dirk-tooth cat (*Smilodon*, top) and a scimitar-tooth cat (*Homotherium*, bottom). Anterior view at left, lateral view at right.

such as some species of *Smilodon* and even of *Barbourofelis*, may have formed groups, as discussed in chapter 4.

Choice of Prey and Ungulate Guilds

Time and again when reading about sabertooths, one comes across the notion that they were specialized in hunting gigantic, thick-skinned prey like mastodons, mammoths, or giant ground sloths, and that their huge canines were the ideal weapons for dealing with such leviathans. It has even been proposed that sabertooth adaptations appeared in the history of land vertebrates each time that herbivores became "megaherbivores"—that is, when they grew to be ten times or more the size of the larger predators (Bakker 1998). As we shall see in later sections of this book, the bulk of anatomical and evolutionary evidence points in a different direction, suggesting that the key to the evolution of sabertooth features was not the killing of much larger prey than that of "normal" or conical-toothed cats, but rather the ability to kill prey within the same size range but doing it in a faster, more efficient fashion, saving energy and minimizing the risk of sustaining injuries during the hunt. In this connection, we should note that in late Permian ecosystems, many of the prey of the giant sabertoothed gorgonopsians were smaller than the predators, and none was ten times bigger. It is also important to bear in mind that, given their size, mammalian megaherbivores are not only extremely difficult or dangerous to hunt, but they are also not the most abundant of potential prey animals in most ecosystems. Medium-sized ungulates like horses and bovids are far more plentiful, and it is reasonable to suppose that, just like today, most large carnivores in the past focused their efforts on taking animals from those species that were more readily available.

2.25. Representative species of the early Pleistocene (top) and middle Pleistocene (bottom) ungulate guilds of East Africa, illustrating dramatic change in species composition in a relatively short time.

Top, from left to right: three-toed horse (*Eurygnathohippus cornelianus*), giraffe (*Giraffa jumae*), spiral horned antelope (*Tragelaphus nakuae*), giant pig (*Kolpochoerus limnetes*), sivatherine giraffid (*Sivatherium maurisium*), impala, (*Aepyceros melampus*), chalicothere (*Ancylotherium hennigi*), hippopotamus (*Hexaprotodon aethiopicus*), and rhinoceros (*Ceratotherium praecox*).

Bottom, from left to right: rhinoceros (*Ceratotherium simum*), giraffe (*Giraffa camelopardalis*), zebra (*Equus koobiforensis*), bovine (*Pelorovis olduvayensis*), alcelaphine antelope (*Megalotragus isaaci*), antilopine antelope (*Antidorcas recki*), reduncine antelope (*Menelekia lyrocera*), pig (*Metridiochoerus andrewsi*), hippopotamus (*Hippopotamus gorgops*), and sivatherine giraffid (*Sivatherium maurisium*).

Being so widespread, many genera of sabertooths have inhabited several continents at the same time, and they have preyed on quite different types of herbivores in different parts of their ranges. Thus the Pleistocene sabertooth *Smilodon* would have fed mainly on bison, horses, and young proboscideans in North America. But when it migrated into South America, it found that bison were absent, horses and proboscideans were somewhat different from the ones in its original range, and there were large ungulates belonging to groups totally unknown in North America, such as the camel-like litopterns and the robust, vaguely hippo-like toxodonts. Faced with such an unfamiliar range of prey, *Smilodon* thrived just as it did in its native continent. Even more extreme was the case of *Homotherium*, which in the Pliocene and Pleistocene ranged from South Africa to England, from Spain to China and South America, and from the Equator to the Arctic. The diversity of antelope that made up a large proportion of its diet in Africa was largely absent in Eurasia and the Americas, where their place was taken by deer, moose, and other ungulates (figure 2.25).

In each case, taxonomic affinities of prey species would be of secondary relevance; the most important factors would be body size, locomotor adaptations, and defensive strategies. Among extant carnivores, the size

of the predator largely determines what size of prey is chosen, and as a general rule, predators from the size of a leopard upward tend to take prey that are their own size or larger. This is a way to ensure that each successful hunt can have an optimal energy return. Apart from this rule, predators choose their prey among the most abundant species in their ecosystem, so that sometimes the prey most commonly taken is somewhat above or below the ideal body size that would fit the predators' energetic needs and hunting abilities. One example can be seen in Kruger National Park in South Africa, where the woodland habitat is home to an abundant population of impala. This medium-sized antelope seems a little too large for the hunting strategies of the cheetah, and a little too small to provide optimum return to a pride of lions, but because it is so abundant, impala is among the most important prey for both cat species, as well as for all the other carnivores in between: leopards, spotted hyenas, and wild dogs. This and other examples show us that we should not be tempted to attribute too specific prey preferences to extinct predators, because this may lead us to imagine too neat schemes of prey partitioning. In fact, all the large carnivores in an area are likely to make the few most abundant ungulate species the primary target, as long as they are not too small or too big and aggressive to hunt.

Sabertooth Diversity

Having occupied such a wide variety of habitats over a vast expanse of geological time, sabertooths have not occupied one single, narrow ecological niche. In several instances more than one species of sabertoothed predators have shared the same environment, and managing to coexist, thanks to subtle differences in body size, locomotor adaptations, and probably prey preferences and hunting styles. For example, the results of biogeochemical analysis of sabertooth fossils from the early Pleistocene site of Venta Micena, in southern Spain, have been interpreted as indicating that the lion-sized scimitar-tooth *Homotherium* preferred to hunt grazing ungulates in the open plains, while the leopard-sized, dirk-tooth *Megantereon* ambushed its browser ungulate prey in the wooded sections of the habitat (Palmqvist et al. 2008). Of course, such subtle responses to environmental factors from vegetation patterns to interspecific competition are part of the dynamic interplay of any predators with their ecosystems. But as we put together more data from the fossil record, we come to an inescapable conclusion: sabertooths played a vital role in past ecosystems, and their spectacular adaptations were not a "one-off," accidental development in nature, but one of the most successful solutions produced by evolution for the problem of being a predator of large prey. Predators are necessarily closely attuned to their habitats. As a result, they have responded to environmental changes with subtle adaptations, which in some instances have been radical enough to lead to the apparition of new species. The natural consequence of these facts has been the amazing diversity of sabertooth species through time, a diversity that we will review in the next chapter.

A Who's Who of Sabertoothed Predators

3

IT DOESN'T TAKE A ZOOLOGIST TO TELL A LIVING LION FROM A tiger, or a leopard from a cougar, but most zoologists would have an embarrassingly difficult time trying to tell apart the bones of any of these pairings. Only a trained specialist, well familiarized with the specific features of big cat skeletons, would be able to pinpoint the subtle differences between the skulls, mandibles, and dentitions of these felids. In the fossil record, where all we have are the bones – often fragmentary – of extinct animals, the features that define a species are not like the ones we use for identifying living animals, features that can be quickly detected by the human eye. Rather, it is often subtle differences in the proportions of the dentition or in the shape of the sutures between bones that tell us what species we are dealing with. With these difficulties in mind, can we hope to create a species-by-species "visual guide" of entirely extinct groups of animals such as the sabertooths? The simple answer is no – but we can produce a reasonable approximation.

This Who's Who is built around life reconstructions (in addition to illustrations of skulls and skeletons), but we cannot hope to reconstruct every named species of sabertooth: many of them are based on fossils that, while presenting clear evidence that they belong to a species different from any other, are not complete enough to permit a reliable restoration. So here we shall select those species that are based on more complete material, and that are morphologically distinctive. Our first goal will be to achieve as clear a picture as possible of the essential features of the larger groups of sabertooths: the felids, nimravids, barbourofelids, creodont sabertooths, marsupial sabertooths, and therapsid pseudo-sabertooths. This may seem simpler than it is: some of the most specialized genera – including *Barbourofelis*, *Smilodon*, and *Thylacosmilus* – were so distinctive in their body and skull proportions that specialists could reasonably bet their money on recognizing the living creature during a hypothetical trip back in time. But if we came across a stocky, leopard-size carnivoran with medium-sized sabers and mandible flanges to match, would we be able to tell at first glance if we were seeing *Hoplophoneus* (a nimravid), *Sansanosmilus* (a barbourofelid), or *Megantereon* (a felid)? Possibly, but just in case, better anesthetize the animal so we can take a close look at the dentition! This implies that, in order to define these groups correctly, a broad visual approach is not enough, and there is no way to avoid getting acquainted with a few more or less obscure anatomical features.

In the introduction to each major group of sabertooths below, I will provide general data on the group's classification, distribution in space and time, and overall appearance. Then I will give more detailed profiles of selected species within each group, choosing those that are better known and/or more characteristic of their respective groups. Other species in addition to the ones appearing in the main profiles may be illustrated and briefly discussed when opportune.

One confusing aspect of dealing with fossil species is the frequent changes in names. The rules of zoological nomenclature require that the first name given to a species in a scientific publication has the priority, and different names later given unwittingly to remains of the same species are called "junior synonyms" and deleted as superfluous. In theory, that approach couldn't sound more reasonable. But imagine that a paleontologist finds a few scraps of something that doesn't quite resemble any known species, and he turns his material into the holotype of a new species, describing it in a local publication. Later, another scientist describes a new fossil that is beautiful and complete fossil, and that seems to be a new species but is actually the same animal discovered by the first scientist. The second scientist gives the fossils a new name, oblivious to the fact that the first scientist had published and named the fragments belonging to the same kind of animal in an obscure journal. Even the editors of the periodical where the new description is submitted fail to notice the previous publication. Decades later somebody accidentally comes across the old publication and discovers the coincidence. In the meantime scientists and even the general public have associated the creature with the name given to the better fossil, but the rules require that the original name be used instead for all future publications.

Claims of priority are not the only reason why scientific names change: the discovery of more complete fossils can provide deeper insights into relationships, and researchers often find that supposedly separate species were just one, or that specimens previously assigned to an existing species belong in new, separate taxa. The scientific names of extant animals also change, especially now that genetic studies produce a better understanding of phylogenetic relationships, but in this case the existence of common names shields the layperson against the confusing effects of such renaming.

As the discovery of sabertooth fossils has proceeded slowly since the early nineteenth century, claims of priority have abounded, and old names have been dusted off time and again. In the following accounts, instead of just giving the current names for the listed species, I mention at least some of the most relevant synonyms, because any curious reader looking into the literature is likely to come across the old names. In addition to that purely practical reason, I have another: some of the stories of changing names show us something about the workings of science and its very human limitations.

As we have seen in chapter 1, therapsids, often known as "mammal-like reptiles," belong, like mammals, to a larger group called the synapsida. Among the features that distinguish synapsids from other reptiles is the trend to develop teeth with different shapes in different regions of the maxilla and mandible; in contrast, most reptiles, including dinosaurs, usually have rows of relatively uniform teeth. In synapsids, the teeth on the anterior part of the maxilla and mandible are often pointed and tightly grouped together like the incisors of mammals. Just behind them appear larger, fang-like teeth called caniniform, and in some therapsids (but not the sabertoothed therapsids), the teeth behind these may develop multiple cusps. This mammal-like regionalization of teeth, known as heterodoncy, is observed even in early pelycosaurs like *Dimetrodon*, and it is a condition that facilitates the development of a sabertooth pattern. Many predatory therapsids displayed respectable canines, so the condition that we see in gorgonopsians was just an exaggeration of an existing trend.

Therapsid sabertooths, also called gorgonopsians, appear some 270 Ma in the late Permian, probably deriving from a group of primitive therapsids known as the biarmosuchids. The first gorgonopsians already show all the synapomorphic characters of the group, such as a well-developed squamosal wing, the presence of a preparietal bone in the skull, a mandible with a high symphysis (chin), and a coronoid process in the dentary bone (Sigogneau-Russell 1970). All the advanced gorgonopsians described here had about five large, pointed incisors on each side of the maxilla and of the mandible. Each of these teeth was oval in section and had a serrated posterior border. Behind the incisors was a huge upper canine and smaller lower one, oval in section and with anterior and posterior cutting edges, both serrated. The skeletons are similar in all these species: robust, but long-limbed for a therapsid, giving the animals a vaguely dog-like stance, although the elbows were somewhat turned outward. The tails were relatively short. These creatures ranged from the size of a coyote to that of a brown bear.

Most of our knowledge of the "gorgons" comes from the fossil sites in the Karoo formation in South Africa, but their fossils have also been found in other African countries and in Russia. They lived only in the late Permian, and none of them appears to have survived the end-Permian mass extinction.

Lycaenops ornatus

Although not larger than a medium-sized dog, *Lycaenops* was a fierce predator, with marked sabertooth features. It had a narrow skull, and the snout was slightly convex dorsally (figure 3.1).

Fossils of this species come from the so-called *Cistecephalus* zone of the Karoo, and the type specimen was a fairly complete skeleton that R. Broom described in 1925, and that was later acquired by the American Museum of Natural History in New York City. At that time, the specimen had not been completely prepared, a task that was undertaken

3.1. Skeleton (top) and reconstructed life appearance of the gorgonopsian *Lycaenops ornatus*. Shoulder height: 40 cm.

in the 1940s when the museum staff decided to mount the skeleton for exhibit. So many relevant anatomical details were revealed by the cleaning and preparation, that in 1948 E. H. Colbert published a complete redescription – one of the outstanding masterpieces of vertebrate paleontology.

Sauroctonus parringtoni

Cranial and skeletal features indicate that this animal was less specialized than other large gorgonopsians. In *Sauroctonus*, the height of the temporal or back region of the skull is broadly similar to the height of the muzzle, while in more specialized gorgons the muzzle becomes higher, in line with the hypertrophy of the anterior dentition. Seen from above, the skull of *Sauroctonus* does not show the great lateral expansion of the temporal region relative to the muzzle that we can see in *Lycaenops* and even more notably in *Rubidgea* (see below). In terms of absolute size, *Sauroctonus* had a total length of about 1.6 meters and was among the larger members of the group, but not nearly the largest. The snout was narrow, with a sloping dorsal profile (figure 3.2).

This species was originally described as *Aelurognathus parringtoni* by the German paleontologist F. von Huene (1950) on the basis of a

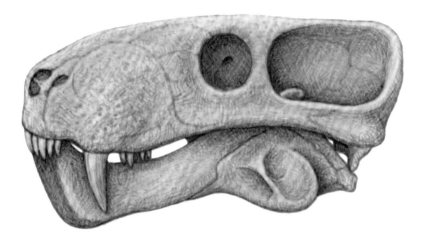

remarkably complete skeleton discovered in the Ruhuhu valley of Tanzania. After a thorough revision of its anatomy, E. Gebauer (2007) relocated Huene's species to the Russian genus *Sauroctonus*, a classification adopted in this book.

Inostrancevia alexandri

Measuring over three meters in total length, this Russian gorgonopsian is one of the largest, but in general morphology *Inostrancevia* is like a scaled-up version of the South African *Lycaenops*. It differs from the latter in having relatively larger upper canines, and the muzzle is also higher compared to the back of the skull (figure 3.3).

Known from relatively complete skeletal remains, *Inostrancevia alexandri* was one of the most spectacular discoveries of the Russian paleontologist V. P. Amilitskii during his exploration of the river valleys of Russia east of the Urals (Battail and Surkov 2000).

Rubidgea atrox

Like *Inostrancevia*, *Rubidgea* could probably reach some three meters in total body length, but its skull, about forty-five centimeters long, was much more robust than that of the Russian gorgon. The snout was narrow and long, but the temporal region of the skull was enormously broadened, so that the cranium was almost as wide as it was long (figure 3.4). In side view, the anterior part of the snout was strikingly high, due to the depth of the maxilla and of the mandibular symphysis, but the sabers were so high-crowned that their tips projected below the ventral outline of the mandible. The body reconstructions of *Rubidgea* shown in this book were built combining the skull anatomy of *R. atrox* with postcranial information from closely related species, such as *Prorubidgea robusta*.

Broom described this species in 1938 on the basis of a well-preserved skull found in the "middle *Cistecephalus* beds" (South African Karroo),

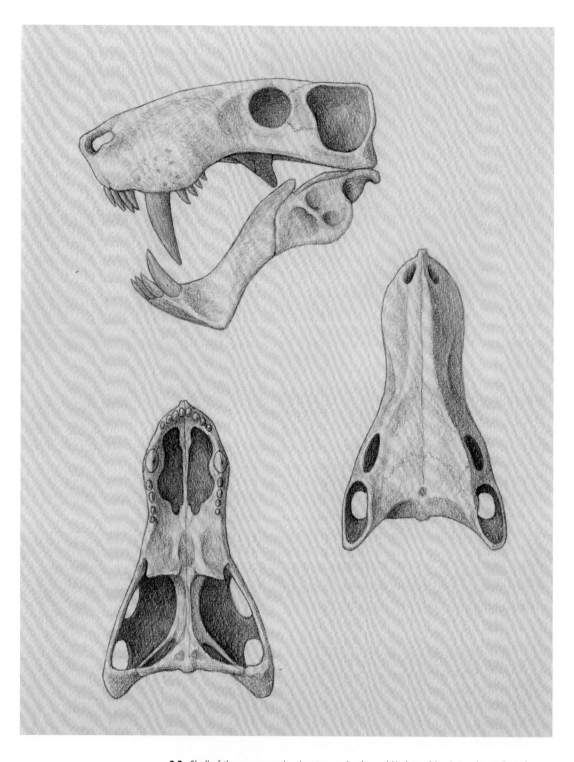

3.3. Skull of the gorgonopsian *Inostrancevia alexandri* in lateral (top), top (center), and ventral views (bottom). In this and the following illustrations of skulls, the dorsal and ventral views are shown without the mandible.

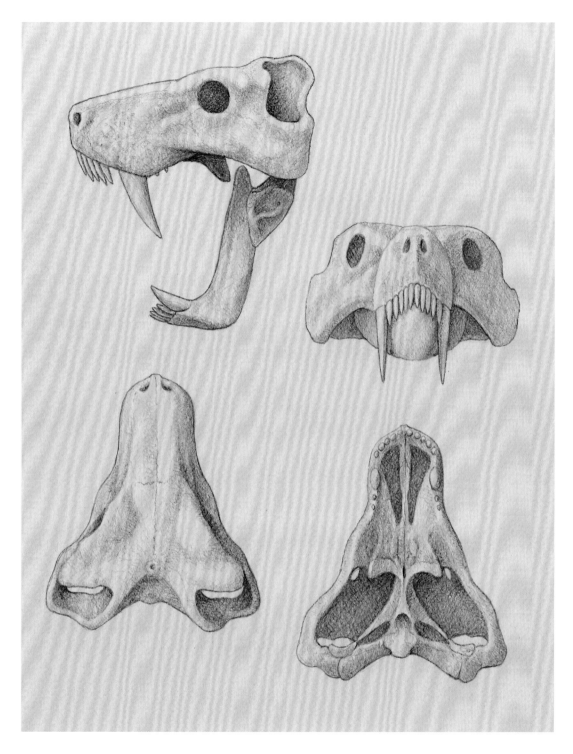

3.4. Skull of the gorgonopsian *Rubidgea* in lateral (top left), frontal (top right), dorsal (bottom left), and ventral (bottom right) views.

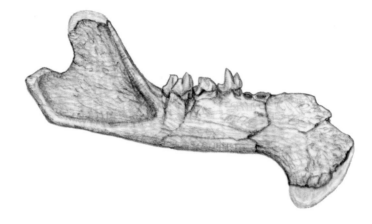

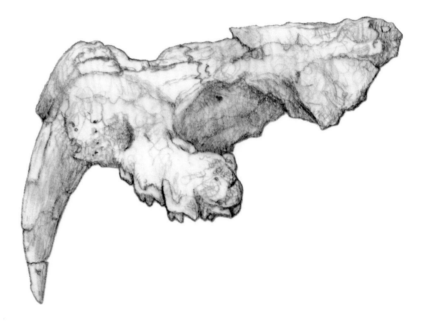

which he called "the finest skull of any South African fossil reptile housed in any museum" (Broom 1938:527).

Marsupial Sabertooths

The spectacular *Thylacosmilus atrox,* which lived in what is now Argentina in the Pliocene, is one of the most extremely specialized mammalian sabertooths, and for many decades it was the only known member of its lineage. At first it was classified in its own subfamily, the Thylacosmilinae, within the Borhyaenidae, a large family of South American predaceous marsupials that included a variety of weasel-like and dog-like forms. Later studies showed that *Thylacosmilus* differed so much from any borhyaenid that the creation of a family Thylacosmilidae was justified (Goin and Pascual 1987), and such differentiation made the absence of any transitional form in the fossil record even

more striking. Only in the 1990s was a second, more primitive thy-lacosmilid genus—*Anachlysctis*—discovered in Miocene deposits from Colombia (Goin 1997), stretching the known record of the family back to some 12 Ma (figure 3.5). More recently, a new genus and species of thylacosmilid, *Patagosmilus goini*, was discovered in Miocene deposits of Argentina; it displays intermediate features between the primitive *Anachlysctis* and the derived *Thylacosmilus*—"derived" in this context meaning that it shows traits not present in more primitive genera (Forasiepi and Carlini 2010).

Thylacosmilids differ from other borhyaenoids in several features of their skulls, including the presence of high-crowned upper canines and a high, ventrally projected mandibular symphysis. They were exclusive to South America, where they apparently went extinct just at the beginning of the main waves of immigration of mammals from North America in the late Pliocene—but before the arrival of placental sabertooths.

3.6. Skeleton (top) and reconstructed life appearance (bottom) of the marsupial sabertooth *Thylacosmilus atrox*. The bones shown in blue are unknown and have been reconstructed on the basis of other borhyaenoid marsupials. Shoulder height: 60 cm.

3.7. Portrait of *Thylacosmilus atrox*. This perspective view allows us to note the narrow head and mandible.

Thylacosmilus atrox was a robust, leopard-sized predator with stocky limbs and a disproportionately large neck and head (figure 3.6). The elongated sabers, with enormous ever-growing roots that extended in an arc up the maxilla and even on top of the orbits, and the elongated mental processes in the mandible, gave this creature a striking appearance, and are evidence of an extreme degree of sabertooth specialization (figure 3.7). Other "machairodont" features of the skull, such as the presence of a postorbital bar and the triangular shape of the skull in ventral view, are also seen in the placental sabertooths of the genus *Barbourofelis* (figure 3.8).

Thylacosmilus was discovered during the 1926 Marshall Field Expedition to Argentina, which found abundant fossils of Pliocene mammals in the province of Catamarca. The marsupial sabertooth fossils were taken to the Field Museum in Chicago, and there E. Riggs (1934) used them as a basis for the original description of *Thylacosmilus atrox*. When Riggs published his study, placental sabertooths had been known to science for about a century, but finding a marsupial that had converged with them in the isolation of South America was quite unexpected.

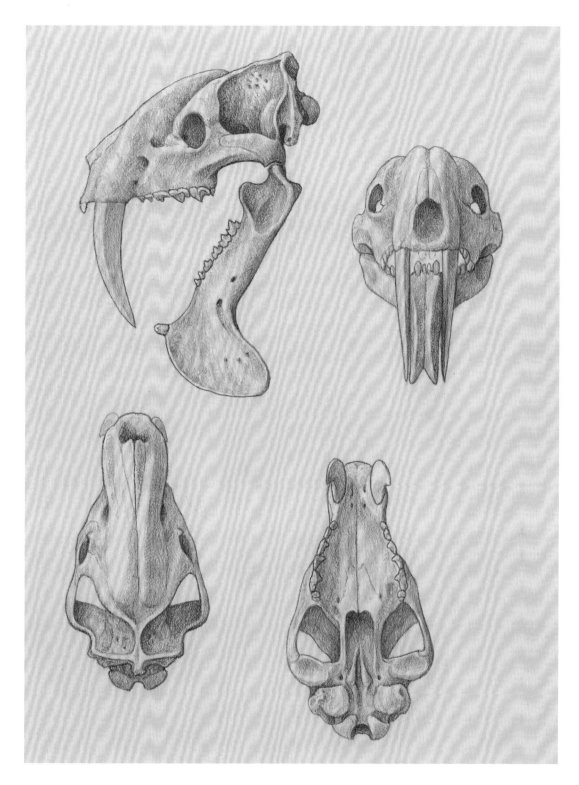

3.8. Skull of *Thylacosmilus atrox* in lateral (top left), frontal (top right), dorsal (bottom left), and ventral (bottom right) views. The premaxilla and upper incisors are unknown and have been reconstructed here on the basis of the presence of wear facets in the internal side of the lower canines.

The Field Museum sample of *Thylacosmilus* fossils lacked parts of the skeleton, but it allowed Riggs to provide a very reasonable picture of the animal, and it remains the most complete collection recovered to date. In the early 1980s, excellent cranial material, as well as a few postcranial bones, was discovered in Buenos Aires and Pampas Provinces, allowing more refined interpretations of the dentition and the biting mechanism.

Creodont
Sabertooths

The order Creodonta contains two families of predatory mammals that were related to, but different from, the true carnivores. These two families are the Hyaenodontidae, a group of vaguely dog-like species, and the Oxyaenidae, which were more stocky and short-limbed, and which would have looked a little like a cross between a cat and an otter. Technically, the most relevant difference between these two groups is the relative position of the carnassial teeth: in the oxyaenids, and in some primitive hyaenodontids as well, the first upper and the second lower molars act as carnassial shears, while in advanced hyaenodontids it is the second upper and the lower third molars that have this function. Creodonts flourished in the Eocene and Oligocene periods, but a few species persisted into the Miocene.

The machaeroidines comprise two genera of creodonts that share the possession of unmistakable sabertooth morphology, but their affinities with either Creodont family have been surprisingly difficult to establish, and there seem to be as many reasons for including them among the hyaenodontids as there are for grouping them with the oxyaenids. In fact, they share with the oxyaenids (and some primitive hyaenodontids) the position of the carnassial shear between the first upper and the second lower molar. M. Dawson and colleagues (1986) classify them as oxyaenids, but M. McKenna and S. Bell in 1997 classify them as hyaenodontids, probably following the criteria of C. Gazin (1946) – essentially, the anatomical similarities with the primitive hyaenodontid subfamily Lymnocyoninae.

Machaeroidines were small to medium-sized predators – that is, from the size of the domestic cat to that of a large lynx – with short legs and a stocky appearance. They appear in the fossil record as full-fledged sabertooths with the early Eocene species *M. simpsoni*, and we know of no transitional forms between them and more primitive creodonts. The only other genus in this family is *Apataelurus*, known from a single mandible attributed to *Apataelurus kayi*, from the Uintan epoch of the middle Eocene, which is the last known representative of the family. Machaeroidines have been found only in North America.

With a reconstructed shoulder height of about twenty-five centimeters, *M. eothen* stood no taller than a large house cat. However, its bones are much more robust, indicating that its body mass was larger, probably like that of a badger, and that it weighed up to ten kilograms (figure 3.9). The elongated skull has many traits of advanced sabertooths, with the maxilla inflated by the massive roots of the canines; a high sagittal

crest; lowered glenoid process; retracted paroccipital process; and large, antero-ventrally expanded mastoid process (figure 3.10). The mandible has a marked mental flange and a reduced coronoid process. The carnassials (first upper molar and lower second molar) of *Machaeroides* are blade-like, almost like those of a cat, and far more evolved toward hypercarnivory than the same teeth in generalized creodonts such as *Lymnocyon*. The upper canines are long and flattened, with a crown height of about three centimeters, but—unlike those of the earlier species *M. simpsoni*—they show no serrations. The lower canines are reduced in size, almost becoming part of the incisor battery. It is clear that the biting mechanism of this diminutive sabertooth was already quite specialized. But its prey would have been small, including fox-sized ancestral horses and dog-sized rhinoceroses. Even so, these herbivores were larger than *Machaeroides*, which would have subdued them thanks to its great muscular strength and forelimbs adapted to grasping.

Described by Matthew in 1909 on the basis of mandible fragments, this species became much better known thanks to the finding of an almost complete skeleton during the Smithsonian expedition to the Bridger Basin of Wyoming in 1940. This specimen was described by Gazin (1946). Fossils of this species have not been found outside that basin and are restricted to the middle Eocene.

3.9. Reconstructed life appearance of the creodont sabertooth *Machaeroides eothen*. Shoulder height: 25 cm.

Nimravid Sabertooths

The family Nimravidae, as defined in recent reviews, includes a series of Eocene and Oligocene catlike animals, ranging from lynx to lion size and found in Eurasia and North America (Bryant 1996b; Peigné 2003; Morlo et al. 2004). Nimravid skulls can be remarkably similar in overall morphology to those of true felids, but they differ in the structure of the auditory region. The septum, or bony wall, that divides the chambers of the tympanic bulla is built from parts of different cranial bones in each.

Some nimravid genera—such as *Nimravus*, *Eofelis*, *Dinaelurictis*, and *Quercylurus*—had only very slight, if any, sabertooth adaptations, and another one, *Dinaelurus*, is best defined as a conical-toothed cat, displaying some cranial similarities with the modern cheetah. The other

nimravid genera–*Dinictis*, *Pogonodon*, *Hoplophoneus*, *Nanosmilus*, and *Eusmilus*–had moderate to extreme machairodont features. As far as we know, nimravids were rather uniform post-cranially, with long bodies and tails, short legs, and short feet that suggest a semiplantigrade or plantigrade posture.

Dinictis felina

Several well-preserved skeletons found in Chadronian deposits of the Rocky Mountain area clearly show the proportions of *Dinictis felina*, which was about the size of a small, female leopard (Scott and Jepsen 1936). Its skull has all the features of sabertooths, although they are only moderately expressed: long, flattened upper canines; enlarged, blade-like caranassials; enlarged mastoid and reduced paroccipital process; and an angular chin (figure 3.11). Its cervical vertebrae, however, were relatively small, with little development of the processes for muscle

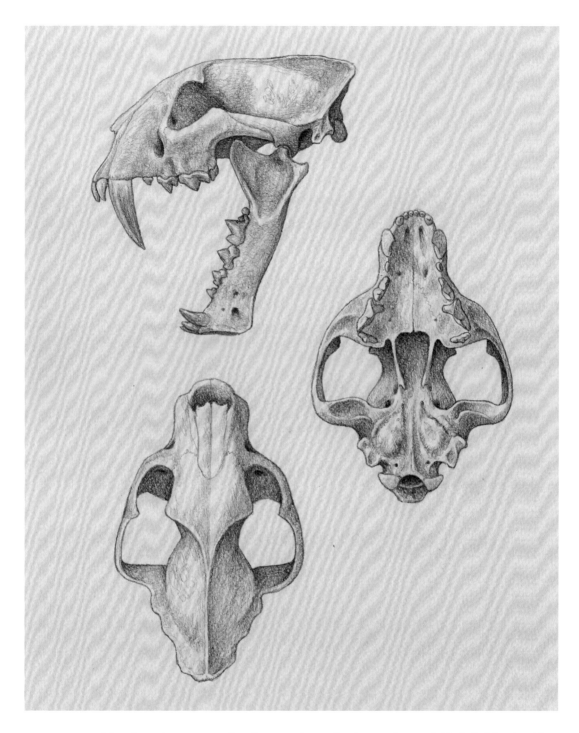

insertions, unlike the robust cervicals of many derived sabertooths. *Dinictis* had relatively gracile, elongate limbs, with the forearm over 90 percent as long as the humerus. But the feet were relatively short, indicating that the posture was probably semi-plantigrade (figure 3.12). With an estimated mass of twenty kilograms, *Dinictis* was more agile than the other Chadron sabertooth, *Hoplophoneus*, but with its short

3.11. Skull of the nimravid sabertooth *Dinictis felina* in lateral (top), ventral (right), and dorsal (bottom) views.

3.12. Skeleton (top) and reconstructed life appearance of *Dinictis* felina. Shoulder height: 45 cm.

feet it would hardly be able to pursue prey over long distances. However, it could have hunted fleeter prey animals, catching them after a short but fast pursuit.

The genus *Dinictis* was established by J. Leidy in 1854, on the basis of a skull housed at the American Museum of Natural History. *Dinictis* is exclusive to North America and ranges from Chadronian (late Eocene) to Withneyan (early Oligocene) times.

Pogonodon platycopis

Closely related to *Dinictis* were the members of the genus *Pogonodon*, relatively large nimravids with more robust dentition than in *Dinictis* and with a larger upper canine (figure 3.13). The species *Pogonodon platycopis* lived in western North America during the Oligocene (Orellan to early Arikareean). Both the genus and species were established by the prolific E. D. Cope (1879, 1880).

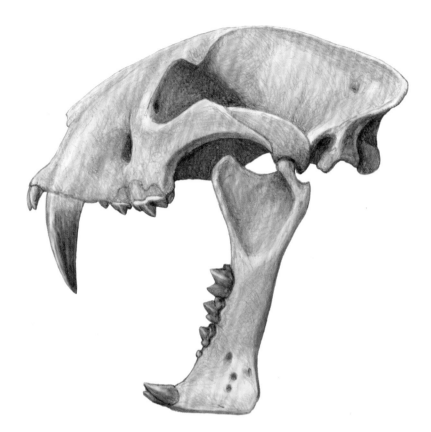

3.13. Skull of *Pogonodon* in lateral view, showing the basic morphology of sabertoothed nimravids.

Genus *Hoplophoneus*

The genus *Hoplophoneus*, established in 1874 by Cope, is exclusive to North America, but it gives its name to the tribe Hoplophoneini, which includes the most derived nimravid sabertooths from the Holarctic. The earliest record of the hoplophoneini corresponds to fragmentary fossils from the middle Eocene of China, too scanty to permit even a clear generic attribution.

HOPLOPHONEUS MENTALIS

H. mentalis is the earliest well-known species of the genus, but it is already quite specialized, more so in some features than the geologically younger species *H. primaevus* (Scott and Jepsen 1936). The skull is that of a highly advanced sabertooth, with very long upper canines, developed mental flanges, hypertrophied mastoids, and retracted paroccipitals, among other traits. The cervical vertebrae are larger than in *Dinictis*, with developed processes for muscle insertions (figure 3.14). *H. mentalis* was not much larger in its linear dimensions (meaning measurements such as body length or shoulder height) than *D. felina*, but its limb bones are more robust, indicating a larger body mass of probably some twenty-five kilograms. *H. mentalis* was evidently adapted to hunt prey larger than itself (figure 3.15).

3.14. Skull and cervical verte-brae (top) and reconstructed life appearance of the head and neck of *Hoplophoneus mentalis*.

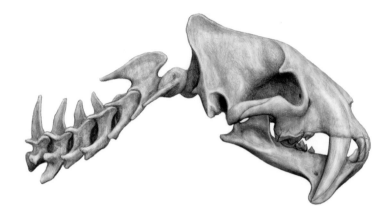

The species *H. mentalis* was described by W. Sinclair in 1921 from fossils from the Chadron formation.

OTHER *HOPLOPHONEUS* SPECIES

The related species *Hoplophoneus occidentalis* (Leidy 1869), known from deposits of Orellan and Withneyan age, is the largest *Hoplophoneus* for which we have reasonably complete skeletons (figure 3.16). With a shoulder height of about sixty centimeters and weighing some sixty kilograms, *H. occidentalis* was a formidable predator, especially for the Oligocene (Riggs 1896).

Another species, *Hoplophoneus primaevus*, was established by Leidy in 1851 and is known from excellent fossils of Orellan to Whitneyan age (figure 3.17).

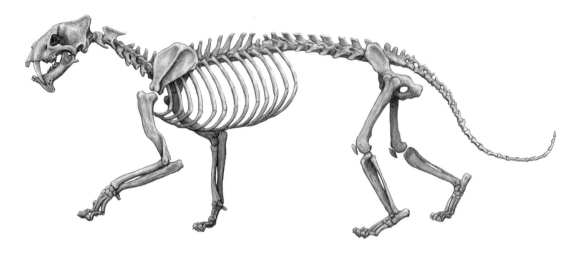

3.15. Skeleton of *Hoplophoneus mentalis*. Shoulder height: 48 cm.

3.16. Reconstructed life appearance of *Hoplophoneus occidentalis*. Shoulder height: 60 cm.

Genus *Eusmilus*

Eusmilus differs from *Hoplophoneus* in having even more extreme sabertooth adaptations. The upper canines are proportionally larger; the carnassials are larger and more blade-like, with a very reduced lingual cusp ("lingual" means the side of the tooth that faces the tongue); and the other cheek teeth are reduced (figure 3.18). The mastoid process has a greater ventral projection, while in the mandible the mental process is larger and the coronoid even more reduced (figure 3.19).

EUSMILUS BIDENTATUS

The species *E. bidentatus* got its name (meaning "two-toothed") from the apparent absence of the first lower incisor, which left only two incisors in each half of the mandible. The original material on which this species

3.17. Skull of *Hoplophoneus primaevus* in lateral view.

was based included some well-preserved cranial and mandibular fossils from the famous Quercy phosphorite sites in France. Unfortunately, the provenance data were scanty, and given the complex nature of the Quercy karstic fillings, it is now impossible to be sure of the age of those fossils. Good fossils of *Eusmilus* failed to appear in Quercy or any other European sites for well over half a century, until the discovery of an almost complete skeleton in Soumailles and a skull in Villebramar, both in France (Ringeade and Michel 1994). The available data show that the animal had body proportions generally similar to those of the species of *Hoplophoneus* (figure 3.20). More recently, the skull of a five-month-old sabertooth kitten, probably belonging to *E. bidentatus*, was found at Itardies (Quercy).

The species *E. bidentatus* was established by M. Filhol in 1872 as *Machaerodus bidentatus*, but the genus name *Eusmilus* was coined by P. Gervais in 1876.

EUSMILUS SICARIUS

This American species of *Eusmilus* was considerably larger than *E. bidentatus*. It is also even more extreme in some of its sabertooth adaptations, especially the development of the upper canines and the flange in the mandible; projecting incisors; reduced coronoid; lowered glenoid; vertical occiput; reduced paroccipital; and an extreme upward rotation of the rostrum, characteristic of the most derived sabertooths (figure 3.21).

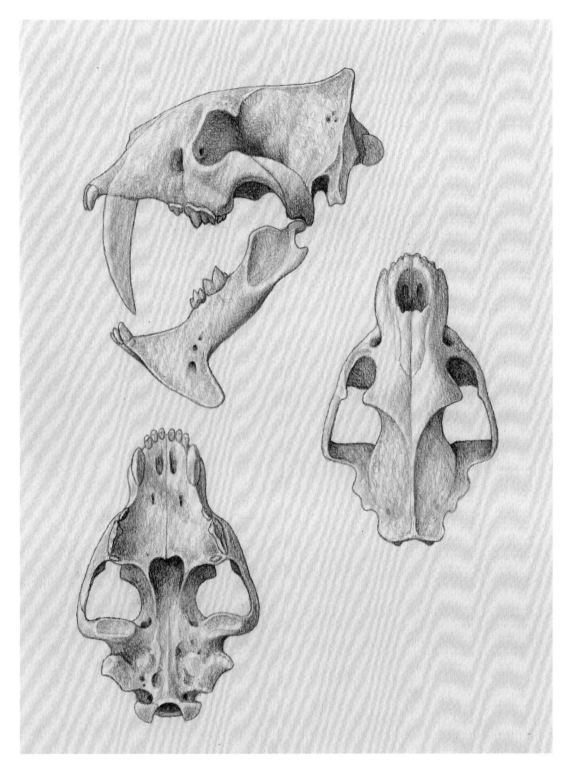

3.18. Skull of *Eusmilus bidentatus* in lateral (top), dorsal (middle), and ventral (bottom) views.

The similarities led W. Sinclair and G. Jepsen (Sinclair and Jepsen 1927; Jepsen 1933) to place it in the European genus. However, other features are more reminiscent of *Hoplophoneus*, and it has been suggested that this is just a member of the *Hoplophoneus* lineage that converged with the European *Eusmilus* (Bryant 1996b). In this book, we keep the traditional generic assignment for lack of a clear solution to this problem (figure 3.22).

OTHER *EUSMILUS* SPECIES

The American species *Eusmilus dakotensis* (Hatcher 1895) is superficially similar to *E. sicarius* but lacks the extreme rotation of the rostrum, and it has been proposed that it is in fact a derived species of *Hoplophoneus* (Bryant 1996b). Also from North America, *Eusmilus cerebralis* is even smaller than *E. bidentatus*, and thus one of the smallest sabertooths known. It is of Whitneyan to Arikareean age and was named by Cope in 1880 from very fragmentary material, but relatively complete skulls have been found since (Bryant 1996b).

Barbourofelid Sabertooths

Barbourofelids were originally introduced to the scientific world with the proposal by C. Schultz and co-authors in 1970 of a tribe Barbourofelini as part of the felid sabertooth subfamily Machairodontinae. Years later, when the status of nimravids as a family different from the Felidae became generally accepted, "barbourofelines" became a subfamily of the nimravids. More recently, it has been proposed that these animals should be in their own family Barbourofelidae, which may actually be closer to the Felidae than to the Nimravidae, and that is the view adopted in this

3.20. Reconstructed life appearance of *Eusmilus bidentatus*. Shoulder height: 48 cm.

book (Morales et al. 2001; Morlo et al. 2004). Barbourofelids were robust, catlike carnivores ranging in size from that of a large lynx to that of a lion. They had short limbs and feet, strong and somewhat rigid backs, and well-muscled necks. The skulls were short and high, especially in the more derived species – which show an extreme development of sabertooth features, including the shortening of the back of the skull; the presence of large mandibular flanges and postorbital bars; verticalized occiputs; and hypertrophied, extremely blade-like carnassials. Other post-canine teeth were reduced or lost. The upper canines had serrations at least on the posterior edge and vertical grooves.

The earliest barbourofelids are known only from mandibular and dental remains attributed to the species *Ginsburgsmilus napakensis*, from the early Miocene of Napak (about 19 Ma) in Uganda (Morales et al. 2001). The first members of this family recorded from Europe also correspond to the lower Miocene; they belong to species of the genus *Prosansanosmilus* (Morlo et al. 2004). The genus *Afrosmilus* is somewhat more advanced than the previous ones, and as its name indicates, it was discovered in Africa (Schmidt-Kittler 1987), but some fossils from the Miocene of Artesilla (about 17 Ma) in Spain have been attributed to a species in this genus, *A. hispanicus*, making *Afrosmilus* the only barbourofelid known from both Africa and Europe (Morales et al. 2001). Those early barbourofelids were about the size of a lynx and had only slight sabertooth adaptations.

Sansanosmilus palmidens

This was a robust, leopard-sized sabertooth, with well-developed machairodont specializations in its skull, but with sabers of only moderate length

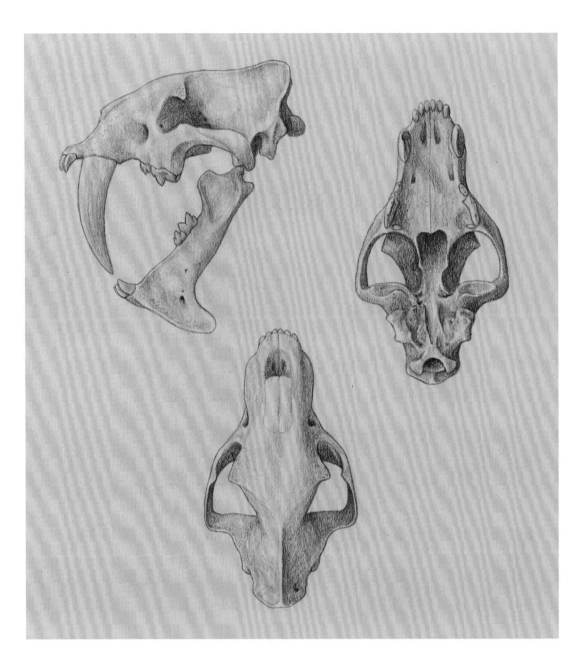

3.21. Skull of *Eusmilus sicarius* in lateral (top left), ventral (right), and dorsal (bottom) views.

(figure 3.23). *Sansanosmilus* differs from its later relative *Barbourofelis* in being smaller, having shorter sabers, less blade-like carnassials, an open orbit, and a less derived mastoid region (figure 3.24). Compared with the contemporary true felid *Pseudaelurus quadridentatus*, *S. palmidens* is relatively shorter-limbed, and its foot bones reveal a more nearly plantigrade posture.

Another species, *S. piveteaui*, known from cranial material from the Sinap formation of Turkey, was at first wrongly classified in the true felid genus *Megantereon*, and then correctly identified as a barbourofelid by D. Geraads and E. Gulec (1997).

Sansanosmilus was discovered at the French site of Sansan, the source of its generic name. It was first described as *Felis palmidens* by H. Blainville in 1841, and like so many other European sabertooths, for many years it bore the genus name *Machaerodus* (an old, alternative spelling to the more familiar *Machairodus*). The name *Sansanosmilus* was coined by M. Kretzoi as early as 1929, but the first detailed description of the rich sample of fossils from Sansan was written by L. Ginsburg in 1961 (1961a).

3.22. Reconstructed life appearance of *Eusmilus sicarius,* yawning.

3.23. Reconstructed life appearance of the barbourofelid *Sansanosmilus palmidens*.

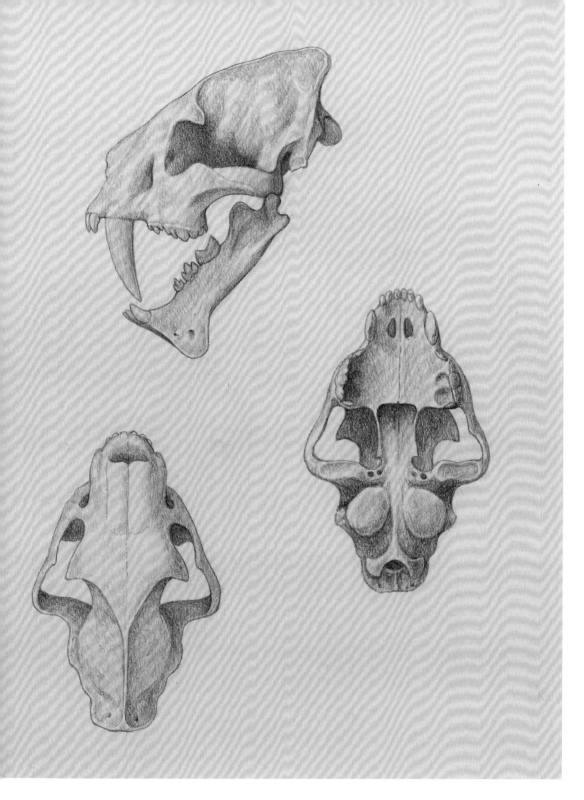

3.24. Skull of *Sansanosmilus palmidens* in lateral (top), ventral (middle), and dorsal (bottom) views.

Albanosmilus jourdani

Closely related to *Sansanosmilus palmidens*, this species was considered for years to belong in the same genus. However, the recent discovery of especially complete fossils has revealed considerable, previously unknown differences with the Sansan species, and it has led to the recovering of the old generic name *Albanosmilus* (Robles et al. in press). This species is known from the Vallès-Penedès basin in Barcelona, Spain.

Genus *Barbourofelis*

The genus *Barbourofelis* was established by Schultz and co-authors in 1970 on the basis of a skull of B. *fricki*, collected in 1947 in Frontier County, Nebraska. The skull was so distinctive that the authors erected a new tribe of sabertooth "cats," the Barbourofelini, to account for its striking features. The defining features of the genus *Barbourofelis* are, among other traits, the presence of a postorbital bar (absent in early barbourofelids such as *Sansanosmilus*); elongated, flattened sabers provided with labial and lingual grooves; and the shortening of the cranium behind the orbits.

BARBOUROFELIS MORRISI

Compared to *Sansanosmilus*, the skull of B. *morrisi* is larger and more specialized (figure 3.25), and it is actually a nicely intermediate form between the relatively primitive European genus and the extremely specialized morphology of the later species *Barbourofelis fricki*. B. *morrisi* was about the size of a large leopard and quite powerful (figure 3.26).

The fossils that led Schultz and co-authors to name this animal a new species in 1970 had been collected by Morris Skinner in 1936, and had remained undescribed in the collections of the American Museum of Natural History for almost thirty-five years. This is quite astonishing when we consider the fact that the holotype skull was a pristine, perfectly preserved specimen that looked unlike any other sabertooth known to that date.

BARBOUROFELIS FRICKI

Barbourofelis fricki was the last member of the Barbourofelidae, and it not only takes all the traits of the family to the extreme, but it is also by far the family's largest member. Its shoulder height would have been around ninety centimeters, and its body weight must have been comparable to that of an African lion. Its short, semiplantigrade limbs display enormous muscular insertions, and its back was shortened, with restricted lateral movements (figure 3.27). Its skull was short and high; its canines were enormously developed, with a huge mandibular flange to match; and its occipital plane was highly verticalized (figure 3.28).

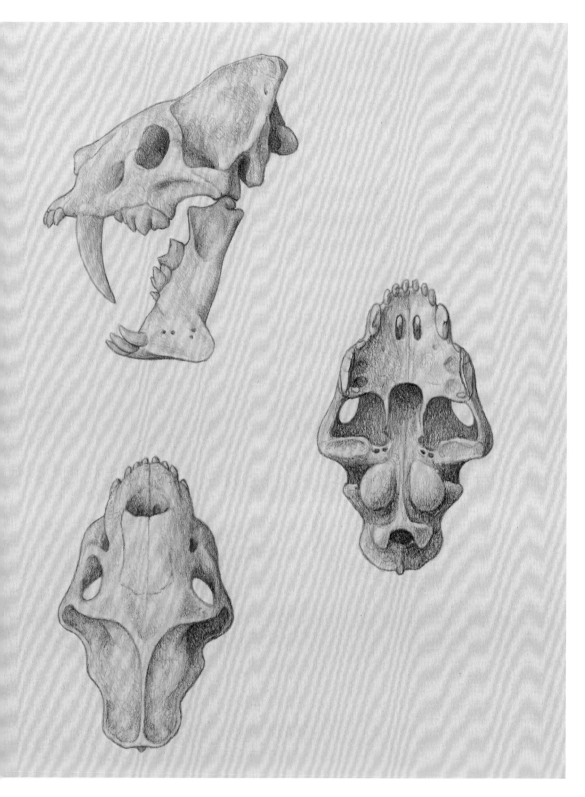

3.25. Skull of *Barbourofelis morrisi* in lateral (top), ventral (middle), and dorsal (bottom) views.

3.26. Reconstructed life appearance of *Barbourofelis morrisi.*

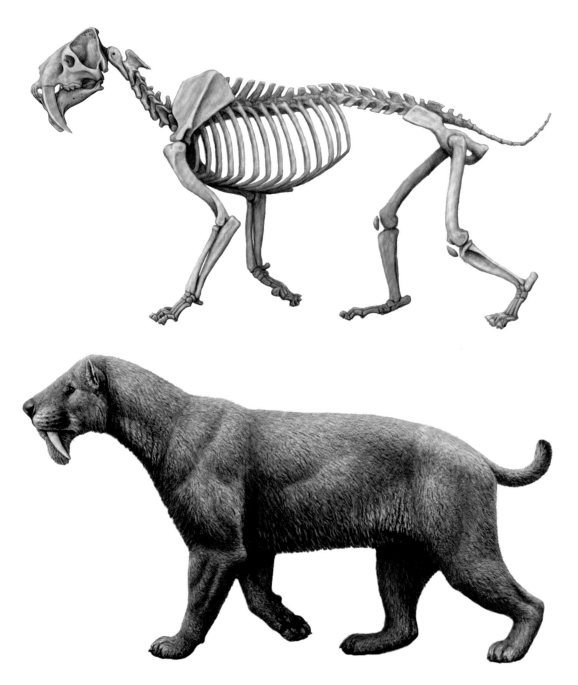

BARBOUROFELIS LOVEORUM

This species is known from a large collection of fossils from the Love Bone Bed in Florida (Baskin 2005). The sample includes unassociated material corresponding to many different individuals, but taken together it nicely complements the picture of the anatomy and proportions of the skeleton of *Barbourofelis* provided by the smaller samples of the other species (see figure 2.14).

3.27. Skeleton (top) and reconstructed life appearance of *Barbourofelis fricki*. The bones shown in blue are unknown and have been reconstructed on the basis of other barbourofelids or closely related carnivorans. Shoulder height: 90 cm.

3.28. Skull and reconstructed life appearance of the head of *Barbourofelis fricki.*

OTHER *BARBOUROFELIS* SPECIES

The American species *Barbourofelis whitfordi* displays several primitive traits that make it resemble the European *Albanosmilus*. The barbourofelid from the Sinap formation of Turkey, variously classified in the genera *Megantereon* and *Sansanosmilus*, actually displays the diagnostic features of *Barbourofelis* and is more likely to belong in that genus, as *Barbourofelis piveteaui* (Robles et al. in press).

Felid Sabertooths: Overview

There are three known groups of felids, or true cats, which developed sabertooth adaptations: these were the metailurins, sometimes called "false sabertooths"; the homotherins, or "scimitar-tooths"; and the smilodontins, or "dirk-tooths" (see chapters 2 and 4 for discussions of these terms). They can be grouped together as three "tribes" within the subfamily Machairodontinae, although the metailurines have been seen by some authors as part of the feline half of the family. Within each of the three machairodontine groups, the earlier species from the Miocene bear the greatest resemblance to modern cats in proportions and morphology, and other than the slightly protruding upper canine tips, the live creatures would have been rather similar to a modern big cat. It is the latest forms of each group, corresponding to the Pliocene and Pleistocene, that show the greatest machairodont specialization, with distinctive skulls, dentitions, and body proportions. In homotherins and smilodontins, there is a clear trend to shorten the lumbar portion of the spine and the tail; and in all three groups, the latest forms show some shortening of the hind limbs. These features, as well as the advanced skull modifications, largely result from convergent evolution, as indicated by the unspecialized anatomies of the earlier species in the three lineages.

The earliest known members of the family Felidae are recorded in the late Oligocene and early Miocene of Europe, with *Proailurus lemanensis* being the best known species. These early cats are sometimes classified in a subfamily of their own, the Proailurinae, and they had a stocky appearance, with short forelimbs and longer hind limbs, as well as semiplantigrade feet with considerable ability for lateral rotation. These anatomical features gave *Proailurus* a remarkable climbing ability, although on the ground it would not have been especially fleet-footed.

In the early Miocene, *Proailurus* or a closely related form gave rise to *Pseudaelurus*, which was conventionally thought to be ancestral to both the modern and machairodontine (or saber-toothed) cats. But it now seems that the genus *Pseudaelurus* as traditionally conceived includes animals that differ too much from each other, so *Pseudaelurus* should be split into at least two separate genera. Smaller species, like the wildcat-sized *"Pseudaelurus" turnauensis* or the large lynx-sized *"Pseudaelurus" lorteti*, look thoroughly feline in morphology, and they are now thought to belong in the genus *Styriofelis*, which is already part of the feline subfamily (Salesa et al. 2012). The leopard-sized *Pseudaelurus quadridentatus*, in contrast, has incipient sabertooth features, such as moderately high-crowned and flattened upper canines and a slightly angular symphysis in the mandible. It thus fits quite well with the ideal ancestral machairodontine, and only minor modifications would have been needed for it to evolve into Miocene felid sabertooths like *Promegantereon*, *Machairodus*, or *Metailurus*.

Among felid sabertooths, the most problematic in terms of classification have been the metailurines, a group with moderate sabertooth features whose members have been seen as intermediate between conical-toothed cats and true machairodonts. An early proposal (Crusafont and Aguirre 1972) had the metailurines classified in their own subfamily, and thus of the same rank as the felines and the machairodontines. Now, however, they are generally considered to be just a tribe—a cluster of closely related genera within the Machairodontinae. But scholars still disagree as to whether they belong with the machairodontines or the felines. In this book we will group them with the machairodonts, because even in the earliest known metailurines several sabertooth features are slight but distinct: all the animals showed more compressed upper canines and more elongated carnassials than modern cats. Nonetheless, they had shorter and less flattened canines than other machairodonts, their mastoid region was little derived, the glenoid was not lowered, and they showed little or no reduction of the mandible's coronoid process. Where known, their cervical vertebrae are hardly longer or more developed than in modern big cats, and their skeletons are generally primitive, with long hind limbs well adapted for leaping. They lived in Africa, Eurasia, and North America, between the Turolian and the early Pleistocene.

Felid Sabertooths: The Metailurins

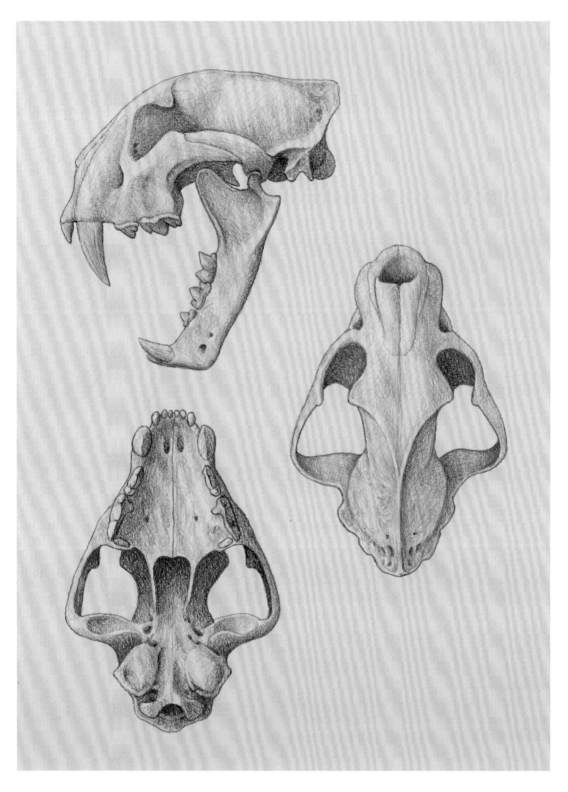

3.29. Skull of the metailurin felid *Metailurus major* in lateral (top), dorsal (middle), and ventral (bottom) views.

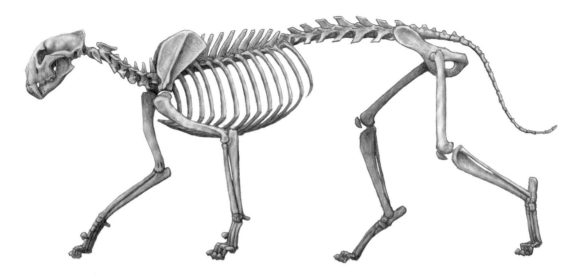

Genus *Metailurus*

3.30. Skeleton of *Metailurus major.* Shoulder height: 73 cm.

This genus was named by O. Zdansky in 1924 on the basis of cranial fossils from China in the Turolian, and later several felid fossils from the classic Turolian localities Samos and Pikermi in Greece were classified in the same genus.

METAILURUS MAJOR

With roughly the body size of a large leopard and an unspecialized morphology, *Metailurus major* looks like a slightly more evolved version of *Pseudaelurus quadridentatus.* The skull and dentition of *Metailurus major* show incipient machairodont features such as moderately long, flattened upper canines; long and narrow premolars; and large carnassials (figure 3.29). The animal was known for over seventy years on the basis of cranial and dental remains only (Zdansky 1924), but recently a remarkably complete skeleton was found in Bulgaria, giving us our first glimpse of the body proportions of this sabertooth (figure 3.30). The Bulgarian skeleton was described by D. Kovatchev (2001) and classified in a new species, *M. anceps*, on the basis of slight differences, although N. Spassov (2002) argues that they seem insufficient to warrant its separation from *M. major*, a point of view that is adopted here. The skeleton of *M. major* corresponds to an animal larger than a male cougar, but whose proportions overall would be very similar except for longer hind limbs (figure 3.31). In contrast, in derived sabertooths like *Smilodon* and *Homotherium*, there is shortening of the lower hind limb bones, to a greater or lesser degree.

METAILURUS PARVULUS

As its Latin name indicates (*parvulus* means small), this species is considerably smaller than *M. major*, with a rather feline-like skull and gracile

3.31. Reconstructed life appearance of *Metailurus major,* leaping.

skeleton. The Chinese species *M. minor* is generally considered to be a junior synonym (that is, an invalid name given to an already properly named taxon) of *M. parvulus.* As in the case of *M. major,* the body proportions of *M. parvulus* were largely unknown for many decades. However, the description of a nearly complete skeleton from Kerassia, Greece, has now been published (Roussiakis et al. 2006). This skeleton reveals body proportions that resemble the cougar in having relatively very long hind limbs, and the snow leopard in the relative gracility of the limb bones (figure 3.32). There is also a wealth of undescribed fossils from the Turolian epoch of China, including several articulated skeletons from Hezheng County, that share the overall morphology and proportions of the Greek specimen while displaying a range of sizes that suggests the presence in China of different populations or even species.

Genus *Dinofelis*

Like *Metailurus,* the genus *Dinofelis* was established by Zdansky in 1924 for Chinese fossils. Ranging from the size of a leopard to that of a small lion, the species of *Dinofelis* are generally larger than *Metailurus,* but nothing precludes their derivation from a species of that genus. Evolution within the genus *Dinofelis* is complex (Werdelin and Lewis 2001), and at least one lineage evolved a morphology converging on that of pantherin cats, with a tiger-like skull and almost conical upper canines. Another

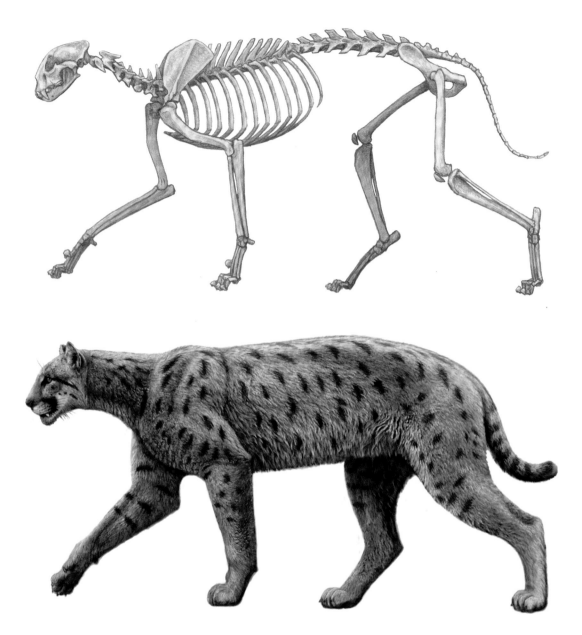

lineage followed the opposite path and evolved more machairodont morphology, with compressed sabers and derived mastoid anatomy.

3.32. Skeleton (top) and reconstructed life appearance of *Metailurus parvulus.* Shoulder height: 58 cm.

DINOFELIS BARLOWI

This species displays the classic features of the genus, with a skull similar in size to that of a jaguar, moderately flattened canines, large carnassials, and a rather primitive mastoid region (figure 3.33).

The first fossils of this species known to science were a damaged skull and an upper canine found at the South African site of Sterkfontein, and described by R. Broom in 1937, but the species was not originally

classified in the genus *Dinofelis*. Indeed, Broom thought that the fossils belonged to a species of *Megantereon* and classified them as *M. barlowi*. In 1955 the leading carnivorist R. F. Ewer restudied Broom's fossils and assigned them to the European genus *Therailurus*, which, as H. Hemmer found a decade later (1965), was just a junior synonym of *Dinofelis* – and thus our animal got its current name.

In 1948 the remains of three individuals were discovered by the University of California African Expedition at the site of Bolt's Farm, also in South Africa, and the blocks of matrix were promptly shipped to

California for preparation. However, none of the material was published until 1991, when H. Cook described the skulls, just a fraction of the collection. This publication provided an excellent picture of the animal's cranial morphology and its variation in this relatively primitive member of the genus. The Bolt's Farm *Dinofelis* post-cranial material remains mostly undescribed, but a cursory overview of the material by L. Werdelin and M. Lewis (2001) and data from fossils from other sites provide a picture of the overall proportions of this animal, which had forelimbs similar to those of a small lion or tiger, but somewhat more robust, and hind limbs proportioned rather like those of a leopard (figure 3.34).

3.34. Reconstructed life appearance of *Dinofelis barlowi.* Shoulder height: 70 cm.

DINOFELIS PIVETEAUI

This is the most sabertooth-like of metailurines, with compressed (if not especially long) upper canines, very elongated carnassials, and an antero-ventrally projected mastoid process (figure 3.35). The neck was not especially elongated or strengthened, but the hind feet were shorter, relative to the fore feet, than in *Dinofelis barlowi,* indicating the same trend toward shortening the hind limb as in other sabertooths.

Ewer described this species in 1955 as *Therailurus piveteaui,* on the basis of a skull from the South African site of Kromdraai. Like all species of the genus *Therailurus,* it was relocated to *Dinofelis* when Hemmer revised the group's taxonomy in 1965.

DINOFELIS CRISTATA

This is the most "pantherine" looking species of *Dinofelis,* and it probably operated much like a small tiger or lion in terms of its hunting habits.

Defined as *Felis cristata* by H. Falconer and P. Cautley in 1836 based on fossils from the Siwaliks in India, the original material was later seen

3.35. Skull and reconstructed life appearance of the head of *Dinofelis piveteaui*.

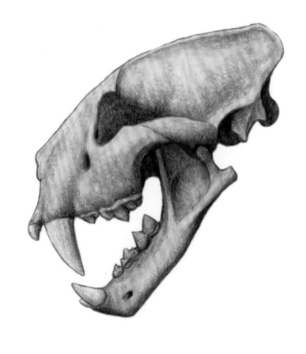

to be identical with *Dinofelis abeli*, a species established by Zdansky (1924) for Chinese fossils. Although Zdansky's genus name is accepted today, the species name coined by Falconer and Cautley had priority and is the valid one.

OTHER *DINOFELIS* SPECIES

In their 2001 revision of the genus, Werdelin and Lewis found a surprising diversity of species, especially in Africa. The typical species from Europe is *Dinofelis diastemata*–known almost exclusively from cranial material, including an excellent skull and mandible from the French Pliocene locality of Perpignan (figure 3.36)–which is probably close to the primitive model for the genus. In Africa the earliest *Dinofelis* comes from Lothagam, but beyond its primitive condition little can be said of the fragmentary material, so it has not been given a species name. A more evolved species is *D. petteri*, from East Africa, intermediate between the Lothagam species and *D. barlowi*. *D. aronoki*, from Kenya and Ethiopia, displays some features of *D. piveteaui* but in a more moderate fashion. *D. darti* from Makapansgat in South Africa is rather similar to *D. barlowi*.

3.36. Reconstructed life appearance of a melanistic *Dinofelis diastemata.*

Other Metailurins

Several genera and species of metailurins are based on material too fragmentary to permit a detailed reconstruction. These include *Stenailurus*, *Adelphailurus*, and *Fortunictis*.

<table>
<tr><td>

Felid Sabertooths: The Homotherins

</td><td>

Paleontologists have realized since the latter part of the nineteenth century that there were at least two contrasting types of sabertooth cats in Europe in the Pliocene and Pleistocene, but the fragmentary nature of most finds made it difficult to define which fossil belonged in what group. Teeth being the most commonly preserved fossils, the shape of the upper canines became the clearest diagnostic criterion, and scientists recognized that one group of sabertooths were characterized by the possession of curved, very flattened sabers with coarsely serrated borders, while the other group had relatively longer, straighter, less flattened canines with smooth edges. These two groups are what we know today as the homotherin, or "scimitar-tooths," and smilodontin, or "dirk-tooths." The discovery in the first part of the twentieth century of complete skeletons of an homotherin cat (*Homotherium latidens*) and a smilodontin cat (*Megantereon cultridens*) in the Villafranchian (late Pliocene to earliest Pleistocene) site of Senèze, in France, allowed paleontologists to define more clearly the characteristics of the two groups. Smilodontins were found to possess robust skeletons with short limbs, while homotherins had relatively long limbs, not too different from those of a modern lion. Homotherins clearly trace their origins to the late Miocene, where species belonging to genera like *Machairodus* and *Amphimachairodus* foreshadow the specializations seen in the Villafranchian sabertooths, while the Miocene roots of the smilodontins are less clear but may lie with *Promegantereon*.

</td></tr>
</table>

Homotherins from the Pliocene and Pleistocene had long forelimbs and shorter hind limbs and backs, which – together with their long necks – make the animals in outline look intermediate between a big feline and a hyena. But the earlier Miocene species retained a more primitive, generalized felid body pattern, with longer hind limbs and backs, which made them more similar, at first sight, to a modern big cat.

Genus *Machairodus*

The genus *Machairodus* includes lion-sized felids from the late Miocene in Eurasia, and possibly Africa and North America. They had flattened, serrated canines, and body proportions somewhat resembling those of a modern tiger, but with a longer and more muscular neck.

The history of the nomenclature of *Machairodus* is inordinately complicated. The characteristic serrated sabers of homotherines had been known to European paleontologists since 1824, when G. Cuvier attributed them to a bear (of all things!) and established the species *Ursus*

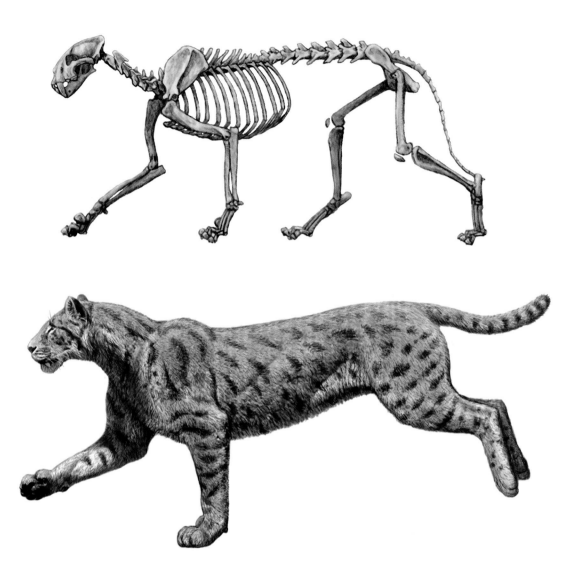

cultridens based on a composite sample of teeth from different species, countries, and geological ages, giving rise to a long series of complications (see the section on *Megantereon cultridens* below). In 1832 the German naturalist J. J. Kaup recognized that those teeth belonged to felids, creating the generic name *Machairodus* and later including Cuvier's animal in it as *M. cultridens.* The name gained acceptance, and by the end of the nineteenth century, almost every sabertooth species known was classified in the genus *Machairodus,* including species that today are attributed to *Sansanosmilus, Paramachaerodus, Megantereon, Homotherium,* or other genera. Many paleontologists, especially French ones, used the word "Machairodus" as a colloquial term, which they applied informally to any sabertooth even if it was perfectly well known to belong to a different genus. The words "machairodont" and "machairodontine" obviously derive from Kaup's popular generic name as well.

In time, and thanks to the discovery of more complete fossils from many sites, the panorama of sabertooth classification began to clarify, and the name *Machairodus* became restricted almost exclusively to the original species *M. aphanistus*, established by Kaup in 1832 for fossils found at the German site of Eppelsheim.

MACHAIRODUS APHANISTUS

The anatomy of this cat has been poorly understood ever since the animal's discovery, due to a very fragmentary fossil record. That situation has changed since the discovery in 1991 of the Cerro de los Batallones fossil sites, where very complete fossils have been found. They reveal *M. aphanistus* to have been a lion-sized felid, with skeletal proportions not unlike those of a modern tiger, although it had a longer and stronger neck and a shorter tail (figure 3.37). Like other members of the Machairodontinae, it had a huge dewclaw that dwarfed the claws of the other digits.

The skull is especially interesting because it displays an unusual combination of primitive and derived features; it appears to be a primitive, "feline-like" skull armed with a set of surprisingly derived sabers (figure 3.38). The typical machairodont features that we listed in chapter 1 (anteroventrally projected mastoid, reduced coronoid, reduced lower canines, protruding incisor battery, and so forth) are either absent or only moderately expressed, but the sabers are high-crowned, laterally flattened, and coarsely serrated – nearly as derived as those in later species like *Amphimachairodus giganteus* or *Homotherium latidens*, which of course have more derived skull features (Antón et al. 2004b). But *M. aphanistus* does show some machairodont features, including a relatively narrow skull, an angular chin, a long post-canine diastema in the mandible, and blade-like carnassials that are much larger than those in a feline of comparable size.

This is a spectacular example of a phenomenon known as mosaic evolution, which occurs when the features that we associate with the end species of a lineage don't evolve simultaneously. In the case of homotherins, the characteristic sabers evolved well ahead of several other features that we consider typical of sabertooths. *Machairodus aphanistus* lacked many of the anatomical refinements that allowed later members of its lineage to use their sabers with greater efficiency, but its enormous success in becoming the dominant carnivore of Vallesian European faunas demonstrates that it was efficient enough to compete successfully with other large predators such as amphicyonid bear-dogs.

OTHER *MACHAIRODUS* SPECIES

Several species of European Miocene sabertooths have been assigned to the genus *Machairodus*, including forms with generally primitive features that look intermediate between the morphology of *Pseudaelurus quadridentatus* and that of *Machairodus aphanistus*. These forms include *Machairodus robinsoni* (Kurtén 1976), *M. pseudaeluroides*

(Schmidt-Kittler 1976), M. *laskarevi* (Sotnikova 1992), and M. *alberdiae* (Ginsburg et al. 1981). *Machairodus kurteni* (Sotnikova 1992) has more advanced features than M. *aphanistus* and resembles the more evolved *Amphimachairodus* and even *Homotherium* in aspects of the arrangement of the incisors, premolar reduction, and loss of lower carnassial metaconid. All these traits actually indicate that this species does not belong in the genus *Machairodus*. A similar assessment can be made of *Machairodus africanus* (Arambourg 1970), based on a well-preserved skull from the Pliocene in Ain Brimba, Tunisia, which–years after its original description–was further prepared and restudied by G. Petter and F. Howell (1987). The skull has many advanced features, such as the arrangement of the incisor alveoli in an arc and the reduction of the protocone in the upper carnassial, that show it to be in an evolutionary grade comparable to or even beyond that of *Amphimachairodus*. Its recent geological age (Villafranchian, late Pliocene) made it the most recent record of the genus, but as in the case of *M. kurteni*, it is evident that this species is not *Machairodus*.

Genus *Nimravides*

The similarity between the genus name *Nimravides* and the family name Nimravidae is a potential source of confusion for the nonspecialist, but *Nimravides* is an undisputed member of the family Felidae. What has been disputed is its affiliation with the machairodontines, and with the homotherins in particular. North America has a good fossil record of "pseudaelurine cats" that seem to grade from primitive, non-sabertoothed forms like *Pseudaelurus intrepidus* to early members of the genus *Nimravides*—like N. *pedionomus*, from some twelve Ma—and then to the later, more clearly sabertoothed N. *thinobates* of ten Ma. Many authors have assumed that all this evolution occurred locally in North America, with some early Miocene *Pseudaelurus* (or *Hyperailurictis*) that migrated from Asia eventually leading to N. *catocopis* as an evolutionary dead end, and then being replaced by another Asian immigrant, the true homotherin *Machairodus coloradensis*. So the theory goes, but several facts complicate this scenario, including the detailed similarity between N. *catocopis* and the Old World *Machairodus aphanistus*. In the 1970s, when the scenario described above was born, hardly anything was known about the cranial or post-cranial anatomy of *M. aphanistus*, since paleontologists had little more than its teeth. It was not even known if *Nimravides*'s upper canines were serrated or not. That state of affairs made it easy to conclude that the observed similarities between advanced *Nimravides* and Vallesian *Machairodus* were due to convergence. In the meantime, a wealth of additional information has come to light, showing that the similarities are so detailed that if a skeleton of N. *catocopis* were found in Batallones, it would be classified as *M. aphanistus* without a second look.

Nimravides catocopis is one of the latest species traditionally attributed to the genus Nimravides, and one of the largest. It attained the size of a large tiger and had long, but powerful, legs and a long back (figure 3.39). The similarities between this species and *Machairodus aphanistus* include the degree of compression of the upper canine, the presence of serrations in its borders, the lack of an upper second premolar, the morphology and arrangement of the incisors, the large lower canines, the development of the coronoid process, and a long list of other features. Furthermore, the differences between the American species N. *catocopis* and *Machairodus coloradensis* are of the same detailed kind as those separating *M. aphanistus* and *Amphimachairodus* in the Old World. All these data suggest the following alternative scenario for the evolution of homotherins in the northern hemisphere: Conceivably, a population of early machairodonts of *P. quadridentatus* grade evolved in Asia and sent successive waves of migrants both to Western Europe and North America. One such wave took the ancestors of *M. aphanistus* to Europe and those

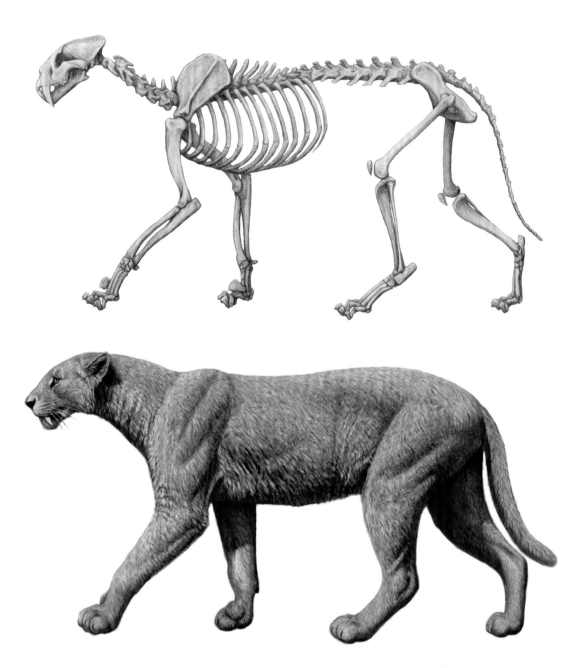

of N. *catocopis* to North America. Later, a more derived homotherin evolved, which became A. *giganteus* in the Old World, and it gave rise to "*M.*" *coloradensis* in America. Positing instead that the ancestors of *Nimravides* and those of *Machairodus* split before the Machairodontinae appeared requires a very detailed convergence and does not seem parsimonious.

The species N. *catocopis* was first named *Machaerodus catocopis* by Cope in 1887, based on a mandibular fragment that included the symphysis. It was Hemphillian (late Miocene) in age.

3.39. Skeleton (top) and reconstructed life appearance of *Nimravides catocopis*. Shoulder height: 100 cm.

Genus *Amphimachairodus*

This was another genus of late Miocene lion-sized machairodontine, which is found in younger deposits than those containing fossils of *Machairodus*. Traditionally thought to be a Eurasian genus, *Amphimachairodus* is now known to have also been present in Africa (Sardella and Werdelin 2007; Werdelin and Sardella 2006) and North America (Martin et al. 2011).

AMPHIMACHAIRODUS GIGANTEUS

The skull of *A. giganteus* differed from that of *M. aphanistus* in having larger, forwardly projected incisors; relatively smaller lower canines; more blade-like carnassials; a more reduced coronid process; and a larger anteroventral projection of the mastoid process (figure 3.40). It was a lion-sized felid with relatively elongated limb bones, but it differed from modern big cats in having a long, extremely muscular neck; a huge dewclaw; and relatively small claws on the other digits (figures 3.41 and 3.42). Unfortunately, no complete skeleton of *A. giganteus* is known, so we have to put together its body proportions by combining fossils from different individuals and sites. Some of the best-preserved skulls come from China (Chang 1957), while a beautiful series of cervical vertebrae was found in Siberia, associated with fragmentary remains of the skull and the rest of the skeleton (Orlov 1936). Partial associations of complete limb bones are known from Greece and Spain (Roussiakis 2002; Morales 1984). Good cranial material was also found in Moldova (Riabinin 1929), and more recently, a beautifully preserved skull, associated with complete forelimb bones, was discovered in Hadjidimovo-1, Bulgaria (Kovatchev 2001). There is no case in which the complete fore limbs and hind limbs of a single individual have been identified, and the lumbar vertebrae and tail are hardly known at all.

Amphimachairodus giganteus ranged throughout Eurasia, from Spain to China and Siberia. In addition, many fossils of late Miocene age from Africa and North America, currently classified in different species, are morphologically very similar and after a thorough revision may show to belong to *A. giganteus*.

Like all the European species of machairodontines that have been known since the nineteenth century, *A. giganteus* has a complicated taxonomical history. For years no clear distinction was made between the Vallesian forms now included in *M. aphanistus* and the more derived Turolian forms. Kretzoi, the Hungarian paleontologist who was notoriously prolific with names for fossil felids, coined the generic name *Amphimachairodus* for the Chinese species *M. palanderi*, on the basis of subtle dental differences with the known European fossils. Half a century later, the Swiss paleontologist G. de Beaumont (1975) revised the Neogene machairodonts and found that all the large Turolian forms shared derived traits in the dentition, separating them from *M. aphanistus*. He proposed

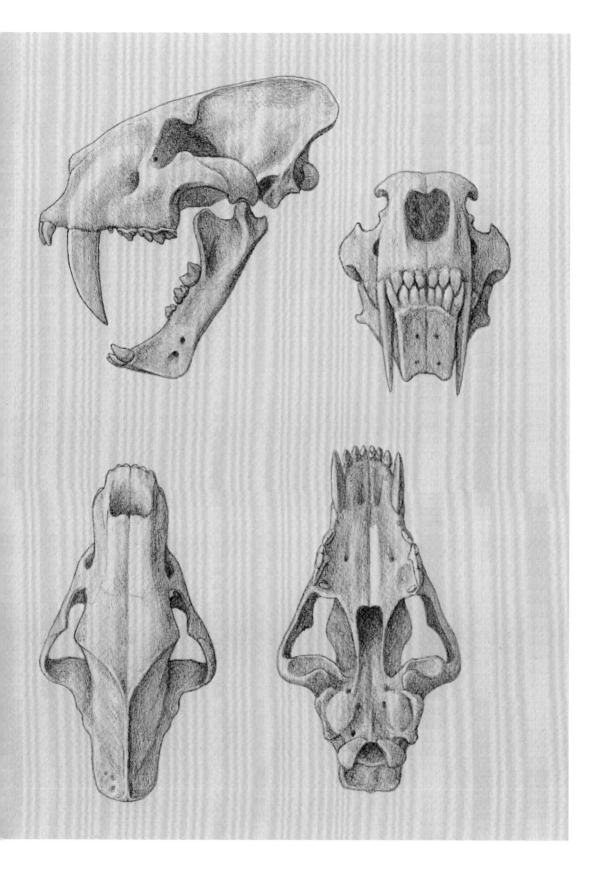

3.41. Reconstructed life appearance of *Amphimachairodus giganteus*. Note the large incisors and small lower canines.

to unite all those forms into a single species, choosing the name *giganteus* – coined by A. Wagner in 1857 for fossils from Pikermi – and including Chinese fossils previously classified as *M. palanderi* and *M. tingii.* Later, Beaumont concluded that the differences between *M. aphanistus* and *M. giganteus* were more than merely specific, and he proposed to group the species into two sub-genera: *Machairodus (Machairodus) aphanistus* and *Machairodus (Amphimachairodus) giganteus,* including the Chinese forms in the latter. Later, the Spanish paleontologist J. Morales (1984), while studying late Turolian fossils from Venta del Moro, found reasons to believe that the differences were of generic rank, so he proposed the classification that we adopt in this book. Most recently, the discovery of the impressive *M. aphanistus* sample from Batallones-1 allowed a clearer assessment of the differences between the two species, fully confirming the generic distinction (Antón et al. 2004b).

AMPHIMACHAIRODUS COLORADENSIS

The timing of machairodontine dispersions to North America seems confusing at first, because *Machairodus* seems to arrive there only some nine Ma, after a long evolutionary history and eventual replacement by *Amphimachairodus* in the Old World. But a review of the anatomy of *Machairodus coloradensis* reveals that it actually has most of the key features that distinguish Old World *Amphimachairodus* from true *Machairodus,* including protruding and enlarged incisors, reduced lower canines, reduced coronoid process, an angular chin, and a derived mastoid anatomy. If we consider *M. coloradensis* as a member of the *Amphimachairodus* group, then its morphology and the time of its arrival to North America are no longer so contradictory (figure 3.43). In 1975 L. Martin and C. Schultz established a new subspecies, "*Machairodus*" *coloradensis tanneri,* based on a mandible from the Kimball formation of Nebraska. More recently, however, Martin and colleagues (2011) have seen this

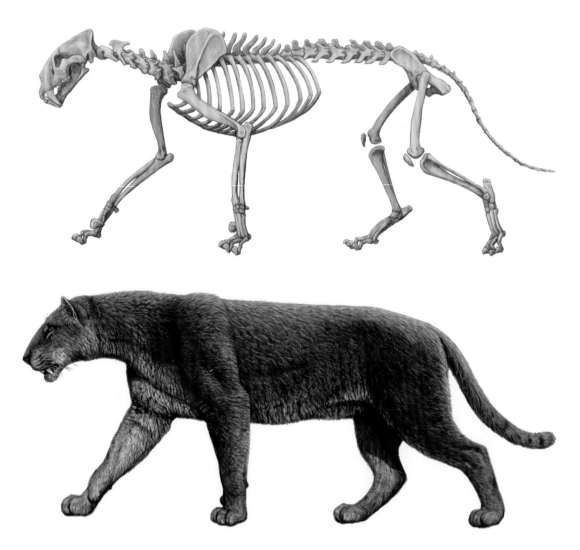

3.43. Skeleton (top) and reconstructed life appearance of *Amphimachairodus coloradensis*.

taxon as deserving the rank of species, and they have attributed it to the genus *Amphimachairodus* as A. *tanneri*.

LOKOTUNJAILURUS EMAGERITUS

As tall at the shoulder as a lioness, *Lokotunjailurus* was more lightly built. In contrast to that of *Homotherium*, in *Lokotunjailurus* the lumbar section of the vertebral column was not greatly shortened (figure 3.44). The holotype skeleton of *Lokotunjailurus* is exceptionally well preserved, including the articulated forepaws, with their claw phalanges in place. This makes evident the disproportionately large size of the dewclaw, larger than the same element in a lion of considerable larger body size, while the claws of the second to the fourth digits were smaller than the same elements in a leopard, which of course is a much smaller animal than *Lokotunjailurus* (figure 3.45). That huge dewclaw would have been a visible feature of the live animal, even when covered with flesh and fur.

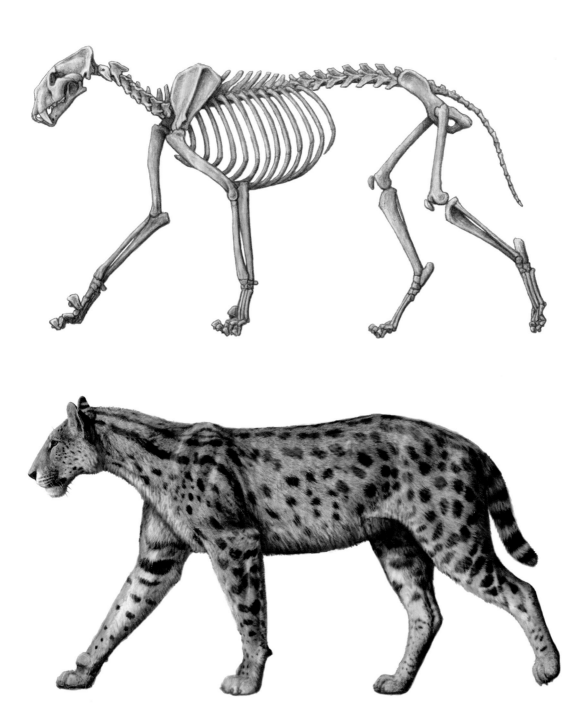

This sabertooth was described by the Swedish paleontologist Werdelin in 2003 from the fossil site of Lothagam in Kenya. The holotype is a nearly complete skeleton first discovered in 1992 by a team led by M. Leakey (Leakey and Harris 2003), although back then it was not known what kind of carnivore the skeleton belonged to. Several bones were found eroding out of a cliff side, and when it was discovered that the specimen continued within the cliff, the team delayed more complete excavation until the next campaign. In 1993 a complex operation

3.44. Skeleton (top) and reconstructed life appearance of *Lokotunjailurus emageritus*. Shoulder height: 90 cm.

3.45. Articulated hand skeleton and reconstructed life appearance of the forepaw of *Lokotunjailurus emageritus*.

finally led to the extraction of a huge block of matrix containing the associated skeleton, which was revealed to be that of a sabertooth cat, at first thought to belong to the genus *Machairodus*. But a detailed study revealed significant differences between the Lothagam sabertooth and known species of *Machairodus*, indicating partial similarities with *Homotherium* and justifying the establishment of a new genus and species.

Genus *Homotherium*

The genus *Homotherium* appears in the fossil record at least 4 Ma, and its origins may be African or Asian since equally old fossils are known

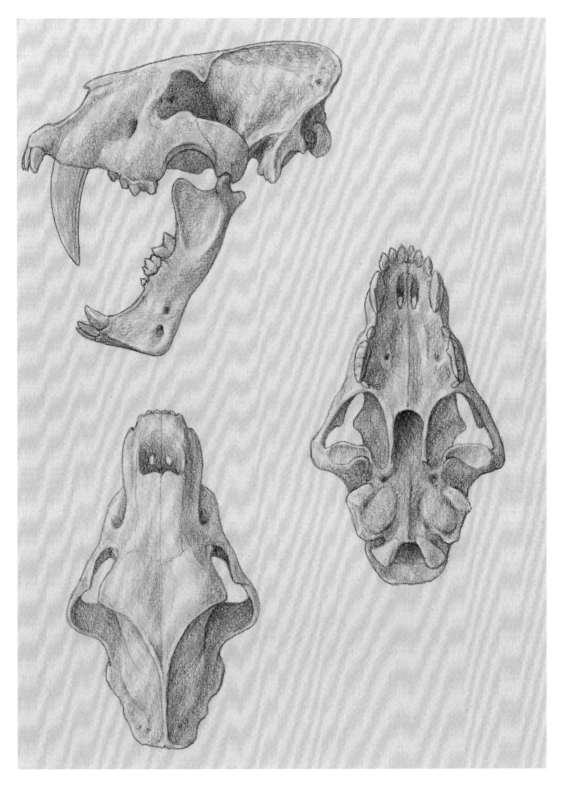

3.46. Skull of *Homotherium latidens* in lateral (top), ventral (middle), and dorsal (bottom) views.

3.47. Skeleton of *Homotherium latidens.* Shoulder height: 110 cm.

from both continents. Even the oldest representatives are already clearly different from their Miocene relatives, such as *Amphimachairodus*.

HOMOTHERIUM LATIDENS

The skull and dentition of *Homotherium latidens* displayed even more profound machairodont adaptations than those of *Amphimachairodus*: it had larger, more protruding incisors; its sabers were more flattened; its carnassials were more blade-like, with total loss of the lingual cusp; the premolars were even more reduced; the coronoid process was lower; and the glenoid process for articulation of the mandible was projected more ventrally (figure 3.46). Its post-cranial anatomy is well known, thanks largely to a nearly complete skeleton from the site of Senèze, in France, and to the large, composite sample from the Spanish site of Incarcal. These and other finds reveal this species to have been a lion-sized sabertooth, with forelimbs slightly more elongated than those of a lion, and with a relatively longer neck and shorter back and tail (figures 3.47 and 3.48). Like all machairodontines, it had strongly muscular forelimbs adapted to handle large prey, armed with a huge dewclaw, but it also had clear adaptations for sustained locomotion on open ground, including a reduction in the size and retractability of the claws. Not only was *H. latidens* relatively lightly built for a sabertooth, but it also was considerably lighter than the Pleistocene lions with which it shared the habitats of middle Pleistocene Europe.

There is a more or less continuous record of this species in Eurasia between the Pliocene (some 3 Ma) and the middle Pleistocene (about 400,000 years ago), when it disappears. After that, the only fossil that has

3.48. Reconstructed life appearance of *Homotherium latidens*, galloping.

been found is a mandible from the North Sea, which carbon-14 dating indicates is about 28,000 years old (Reumer et al. 2003). It is difficult to infer from such an isolated find that this animal had been present in Europe all the time since its apparent extinction. It is also possible that the find represents a punctual immigration event from North America, where *Homotherium* is present uninterruptedly from the Pliocene to the end of the Pleistocene.

The first teeth of *Homotherium* known to modern science were found in a British cave called Kent's Hole and were described in 1846 by R. Owen (the scientist who coined the term "dinosaur") under the name *Machairodus latidens*. Almost half a century later, the Italian paleontologist E. Fabrini set to study the Villafranchian machairodonts of Tuscany, in Italy, and in 1890 he created the generic name *Homotherium* for two of the larger species he recognized there, which he named *H. crenatidens* and *H. nestianum*. But Fabrini's name did not become very popular, and it was hardly used until C. Arambourg applied it in 1947 to specimens discovered in Ethiopia. In 1954 J. Viret used it for fossils from the French site of Saint-Vallier, further proposing to label much of the Villafranchian material from Europe *H. crenatidens*. During the twentieth century, the taxonomic history of *Homotherium* has been very complex, with new species being named on the basis of subtle, often invalid, differences in dental morphology and size. Current views tend to recognize a single species, *Homotherium latidens*, in the Pliocene and Pleistocene of Eurasia (Antón et al. 2005, 2009).

HOMOTHERIUM SERUM

This species is found in the late Pleistocene of North America, and it differs from the Old World species H. *latidens* in a series of subtle morphological features. It had a shorter diastema (a term meaning the empty space between two teeth, in this case the third incisor and upper canine), there is a distinct "pocketing" in the anterior margin of the masseteric fossa of the mandible, and the forehead seems to be wider in dorsal view. The sabers were not especially large, but they were very flattened (figure 3.49). The hind limbs were even shorter relative to the forelimbs, giving the animal a more sloping back (figure 3.50).

The species *Homotherium serum* was established by Cope in 1893 as *Dinobastis serus*, but later scholars did not find it justifiable to keep the animal in a genus separate from *Homotherium.*

The best known sample of this species comes from Friesenhahn Cave in Texas, which has yielded the remains of several individuals, including articulated skeletons of adult and young animals. This exceptional find allows us to reconstruct the body proportions of *Homotherium* cubs and provides insights into the eruption sequence of its dentition (Rawn-Schatzinger 1992).

OTHER *HOMOTHERIUM* SPECIES

An early species of *Homotherium*, with more primitive features than those of *H. serum*, inhabited North America in the Blancan (Pliocene), and it has recently been redescribed as *Homotherium ischyrus* (Hearst et al. 2011) on the basis of a nearly complete skeleton from Birch Creek, Idaho. In general morphology the animal is similar to the Old World species *Homotherium latidens*, but some characters—such as the retention of a two-rooted third upper premolar in the mandible and an elongated lumbar section in the vertebral column—suggest an early separation from the European lineage.

One of the most relevant findings of recent years regarding the distribution of *Homotherium* has been the discovery of excellent material of this genus in South America. The fossils come from a tar seep deposit in northeastern Venezuela and have been classified by its discoverers in a new species, *Homotherium venezuelensis* (Rincón et al. 2011). This species is of early to middle Pleistocene age and shares many features with the Old World *H. latidens* (figure 3.51).

Homotherium finds are relatively common in Africa, but because most of them are fragmentary, it is difficult to be sure about the specific identity of most specimens. On the basis of one reasonably complete

3.50. Reconstructed life appearance of an adult female *Homotherium serum* with cubs on a snowy hillside.

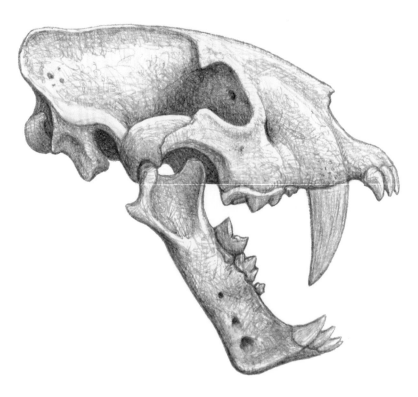

3.51. Skull of *Homotherium venezuelensis.*

skull from Pliocene deposits in Afar, Ethiopia, Petter and Howell (1988) erected a new species, which they named *H. hadarensis* (figure 3.52).

Xenosmilus hodsonae

This is the only known species in the genus *Xenosmilus*, and its morphology came as a surprise for its discoverers because it was so unexpected for a member of the Homotheriini. It differed from other members of its tribe in a series of details in the skull and dentition, but most conspicuously in its body proportions. Its robust and shortened limbs resemble those of the smilodontin dirk-tooths (see below) more than those of its homotherin cousins (figure 3.53). As in *Homotherium*, the canines are flattened and coarsely serrated, the incisors are very large and procumbent, the premolars are very reduced, and the carnassials are huge and blade-like (figure 3.54). Unlike in *Homotherium*, the forehead is narrow, and the constriction behind the orbits is very marked. The diastema between the third incisor and upper canine is reduced, so the upper incisors and canines could operate more as a unit during the bite (figure 3.55). This feature has led the authors of the original description of the species (Martin et al. 2000, 2011) to coin the term "cookie-cutter cat" to define this particular craniodental morphology, which in their view implied a different way of dealing with the animal's prey (see chapters 2 and 4).

This genus and species is clearly represented only at the site of Haile 21A in Florida, where relatively complete skeletons of two individuals were found (Martin et al. 2000). A fossil tentatively identified

3.52. Skull and reconstructed life appearance of the head of *Homotherium hadarensis*. The original fossil skull has a slight crushing that has been corrected in this drawing.

as *Xenosmilus* was found in Uruguay (Mones and Rinderknecht 2004), but it is too fragmentary for positive identification and may equally well correspond to *Homotherium* (Rincón et al. 2011).

Felid Sabertooths: The Smilodontins

Smilodontin cats can be considered in some ways to be the ultimate machairodonts, as they include the giant ice age species that all of us have come to identify with the idea of a sabertooth. Members of the genus *Smilodon* were huge, enormously strong creatures, and they had the most impressive upper canines of any sabertooth. But in many ways they were less specialized than other species that we have already discussed. In terms of the evolution of their dentition, smilodontines were

3.53. Reconstructed life appearance of *Xenosmilus hodsonae*, sitting. Shoulder height: 100 cm.

less precocious than their homotherin cousins, so the carnassials of *Promegantereon* and even those of *Megantereon* are relatively primitive and resemble those of a "normal cat" in being relatively short and only moderately flattened, and in retaining a sizable protocone, or internal cusp. In contrast, the carnassials of late Miocene homotherines were already more blade-like, elongated, and flattened, and their evolution rapidly reduced the protocone. The flattening of the upper canines was also greater in even the earliest homotherines than it ever became in the smilodontines. In terms of cranial adaptations, derived barbourofelids were far more specialized than even the latest smilodontines, whose skulls always retained an overall catlike air.

Promegantereon ogygia

This animal was about the size of a small leopard and rather similar to it in general proportions, although with a longer neck and shorter tail (figure 3.56). Cranially, it had slight but unmistakably machairodont features, including elongated, flattened upper canines; an angular chin; and a large mastoid process (figures 3.57 and 3.58).

The species was established (as *Felis ogygia*) by Kaup in 1832, based on some mandibular fragments found at the German site of Eppelsheim,

of Vallesian (late Miocene) age. In 1913 the British paleontologist G. E. Pilgrim judged that the Eppelsheim remains did not correspond to any member of the extant genus *Felis*, but rather to a sabertooth cat, so he included them in his new genus *Paramachaerodus*, coined by him for several fossils of leopard-sized felids of later (Turolian) age, including *Machairodus schlosseri* from Pikermi and *M. orientalis* from Maragha. In 1938 Kretzoi coined the genus name *Promegantereon* for the Eppelsheim fossils, and in 1975 Beaumont revised them and found close similarities between *P. ogygia* and the middle Miocene *Pseudaelurus quadridentatus*. He also thought that there might be a direct phyletic relation between *P. ogygia* and the younger *P. orientalis*. Current scholarly opinion is that the primitive features of *P. ogygia* (which led Beaumont to see affinities with *Pseudaleurus*) are real enough to separate it from *P. orientalis* and keep it in a separate genus, for which Kretzoi's name *Promegantereon* is available.

This is only a summary of the systematic complications that *P. ogygia* has been subject to since the early nineteenth century, but during all this time the handful of mandibular fragments from Eppelsheim were virtually the only fossils known from this species—it often seems that the more

3.54. Reconstructed life appearance of *Xenosmilus hodsonae,* snarling.

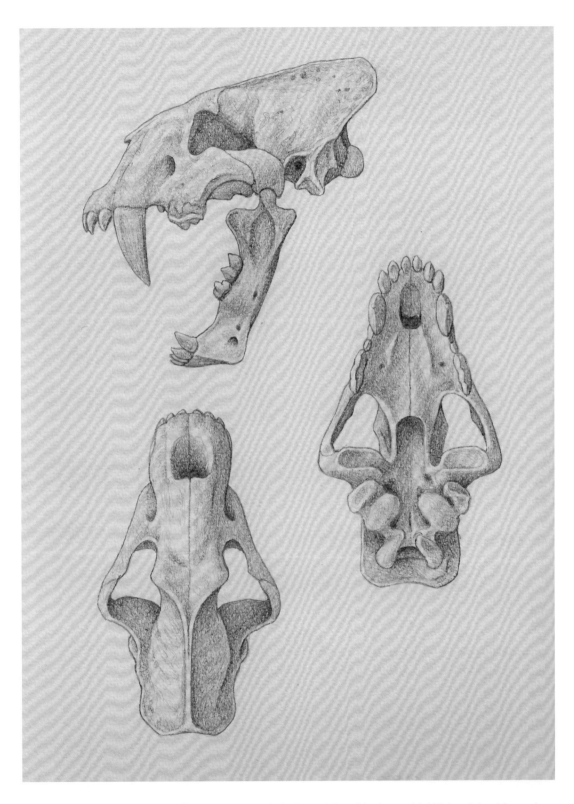

3.55. Skull of *Xenosmilus hodsonae* in lateral (top), ventral (middle), and dorsal (bottom) views.

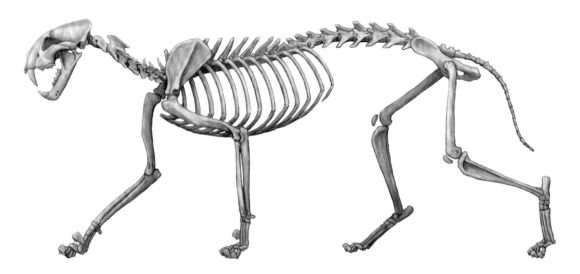

3.56. Skeleton of *Prome-ganteron ogygia*. Shoulder height: 60 cm.

fragmentary the fossils, the greater the amounts of ink that are spilled to discuss their affinities! It was only in the early 1990s, with the discovery of the Cerro de los Batallones fossil sites in Madrid, that the anatomy of this cat finally became well known. The Spanish paleontologist M. Salesa wrote his PhD thesis on the Batallones *P. ogygia*, producing an in-depth view of this previously mysterious cat (Salesa 2002).

Genus *Paramachaerodus*

The genus *Paramachaerodus* (Pilgrim 1913) comprises species of leopard-sized sabertooth cats that lived in Eurasia during the late Miocene.

PARAMACHAERODUS ORIENTALIS

This Turolian-aged smilodontin is less well known than *P. ogygia*, but it is broadly similar to it in its known features, although it was marginally larger and showed slight serrations on its sabers (figure 3.59). B. Kurtén (1968) considered it to be so similar to the *Megantereon* of the Pliocene and Pleistocene that he included it in the same genus as *Megantereon orientalis*, but later scholars have not favored that classification (Salesa et al. 2010a).

PARAMACHAERODUS MAXIMILIANI

Described by Zdansky in 1924 on the basis of cranial material from the Turolian of China, this species is distinguished from *P. orientalis* by its larger size and the very derived morphology of its sabers, which are flattened, have serrated edges, and resemble a miniature version of the canines of an homotherin cat more than those of *P. orientalis*.

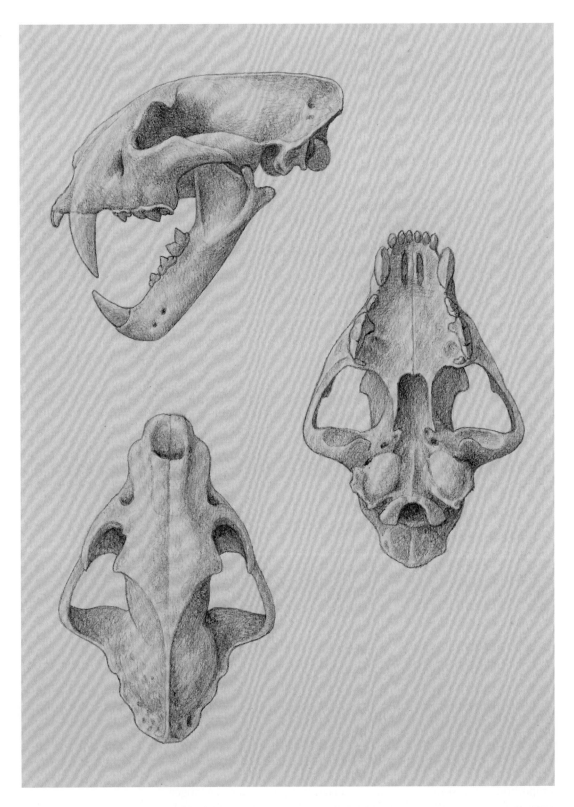

3.57. Skull of *Promegantereon ogygia* in lateral (top), ventral (middle), and dorsal (bottom) views.

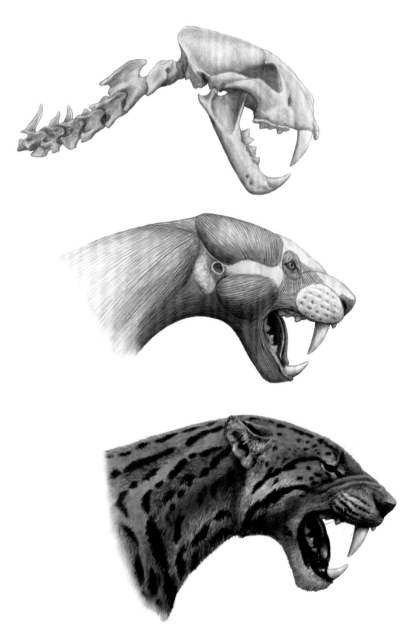

Genus *Megantereon*

The genus *Megantereon* comprised species of leopard- to jaguar-sized sabertooths that lived between the early Pliocene and the middle Pleistocene in Africa, Eurasia, and North America. Just as many naturalists consider the leopard to be the ultimate big cat, I find *Megantereon* to be the ultimate sabertooth. Less imposing than the hyper-robust *Smilodon*, *Megantereon* struck a balance between strength and grace. In size and proportions it was similar to the modern jaguar, although it had a longer neck and a shorter tail. Like the jaguar, it could explode out of concealment with lightning speed, and it was large enough to hunt big prey such

3.59. Skull of *Paramachaerodus orientalis*. Sections shown in blue are unknown and have been reconstructed on the basis of closely related species.

as horses and deer but still agile enough to be a proficient climber. Its skull and neck displayed all the adaptations of a derived sabertooth, and it would have been swift and efficient in killing its prey. This combination made it enormously successful, allowing it to spread from South Africa to Greece, and from Spain to China and to North America. The history of the classification of *Megantereon* is incredibly complicated for an animal that appears so unmistakable to us. When Cuvier described the first collection of machairodontine teeth in 1824, he used a combined sample that included two teeth of *Megantereon* from the Pliocene of Valdarno, Italy, and one of *Machairodus*, from the Miocene of Eppelsheim, Germany. He believed that all the teeth belonged to a single species of animal, which he amazingly considered to be a bear, and he named the species *Ursus cultridens*. Cuvier is often quoted as having said "give me a tooth and I will reconstruct the entire animal," but it seems he went a little too far on this occasion. Yet surely he could not suspect what a mess he would cause with his decision of creating a new bear species from teeth of animals spanning two countries and some 7 million years! Four years later, the French paleontologists Croizet and Jobert described a mandible of *Megantereon* from the French site of Les Etouaires, and they correctly recognized it as felid, giving it the name *Felis megantereon*. But when they found the upper canine of the same kind of animal in the site, they saw the similarities with Cuvier's species, and instead of associating the canine with the mandible, they classified it as *Ursus cultridens*.

We must bear in mind that paleontologists in the early nineteenth century had never seen the combination of a flattened, elongated canine with a catlike skull and mandible, and they were understandably confused. But in the same year of 1828, the puzzle was solved when the French paleontologist M. Bravard identified a skull of *Megantereon* with its canines in place in the French site of Mont Perrier, and he further recognized that the mandible described by Croizet and Jobert belonged to the very same creature, for which Bravard proposed the name *Megantereon megantereon*. Since the canines of *Megantereon* bear no serrations, Bravard proposed to use that feature to distinguish it from the other kind of sabertooth, exemplified by the crenulated upper canine from Eppelsheim that Cuvier had used (in part) to create his *Ursus cultridens* (which we now know belonged to *Machairodus aphanistus*). Bravard judiciously proposed that Cuvier's specific name *cultridens* should be kept, but preceded by the generic name *Machairodus* (coined by Kaup), so from then on, the sabertooth cats with serrated upper canines should be called *Machairodus cultridens*. But, alas, Cuvier had based his species on the teeth of both *Machairodus* and *Megantereon*, so Bravard's attempt at clarification was doomed to failure.

In his 1890 revision of the classification of machairodonts from Tuscany (Italy), Fabrini grouped the specimens with non-crenulated canines under the name *Machairodus (Meganthereon) cultridens* (yes, some scholars of old spelled the name *Meganthereon*) acknowledging the first two-thirds of Cuvier's original sample. But in 1901 the French paleontologist M. Boule reviewed the European machairodonts, and he considered *Machairodus cultridens* to be a valid species, for which *Machairodus megantereon* would be only an invalid, junior synonym! This taxonomic nightmare persisted until, in 1979, the Italian paleontologist G. Ficcarelli carried out a new revision of the Tuscan machairodonts, carefully applying the rules of nomenclature and concluding that the name *Megantereon cultridens* had priority over all the others, to define the panther-sized sabertooths with non-crenulated canines.

MEGANTEREON CULTRIDENS

This was a jaguar-sized smilodontin, characterized by short, robust limbs; a very long and muscular neck; and a skull with more marked machairodont specializations than those of *Promegantereon* or *Paramachairodus* (figure 3.60). The general morphology of the skull is remarkably similar to that of the Eocene nimravid *Hoplophoneus*, including a comparable development of the upper canine, mandibular flange, reduction of the coronoid process in the mandible, and development of the mastoid process. This is a very detailed convergence between two genera belonging to different carnivoran families and separated by more than 30 million years (figure 3.61).

It was not until a century after Cuvier described the teeth of *Megantereon* for the first time that the post-cranial anatomy of this animal became

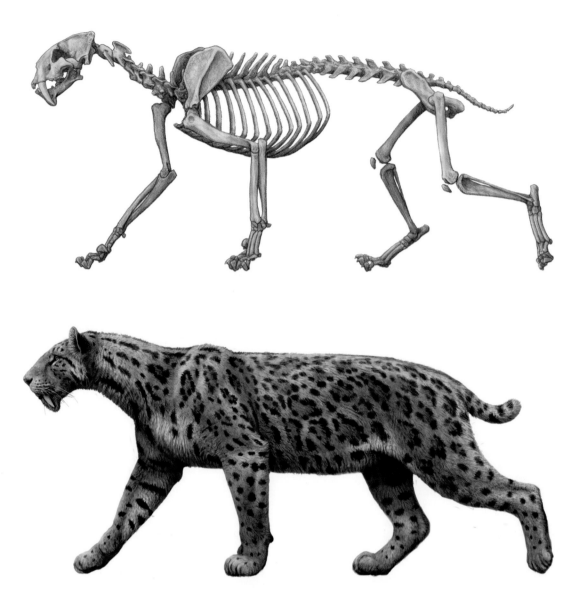

3.60. Skeleton (top) and reconstructed life appearance of *Megantereon cultridens*. Shoulder height: 70 cm.

well known, thanks to a nearly complete skeleton found at the French site of Senèze, and preliminarily described by the Swiss paleontologist S. Schaub in 1925. Schaub defined the animal as a jaguar-sized cat with short, robust limbs that were in marked contrast to the more gracile proportions of the other Villafranchian machairodont, *Homotherium latidens*. Schaub planned to write a detailed monograph about the Senèze skeleton, but he died without completing it. Thus, his preliminary description was the only one published for many decades, while the skeleton remained on exhibit in its glass case at the Museum of Basel, Switzerland. It was only in 2007 that the Scandinavian paleontologists J. Adolfssen and P. Christiansen published an updated, detailed description of that skeleton.

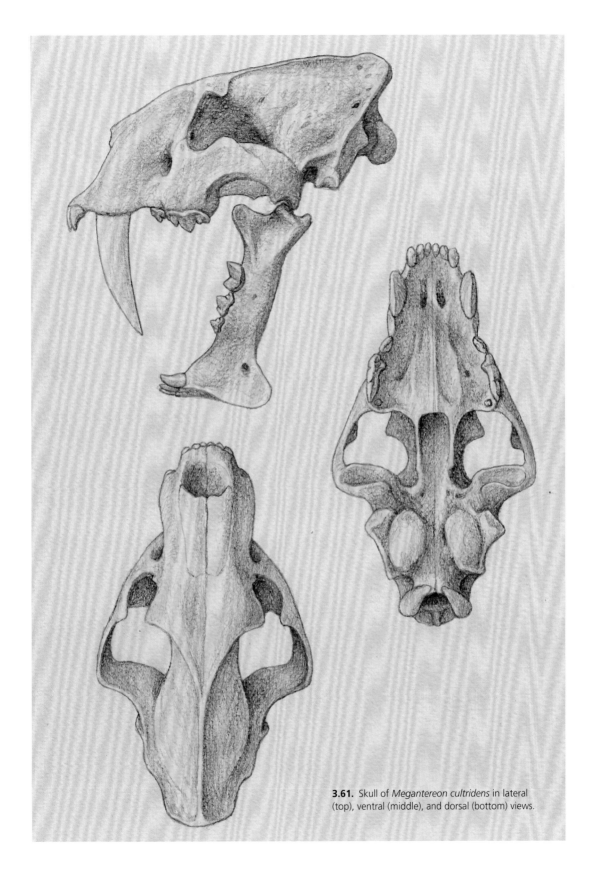

3.61. Skull of *Megantereon cultridens* in lateral (top), ventral (middle), and dorsal (bottom) views.

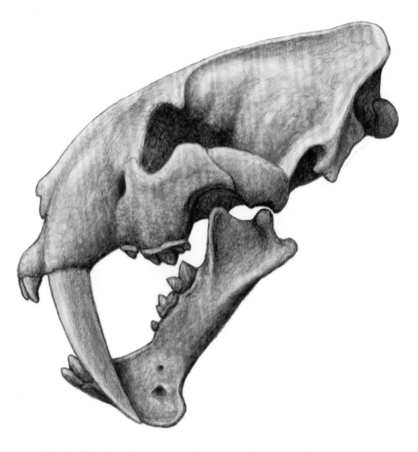

The skull of the Senèze specimen was illustrated many times, and it became the standard image for the cranial morphology of the species. However, it had a rather unexpected feature: the occiput was rather elevated as in the most derived smilodontine species, *Smilodon populator*, instead of being inclined as it seemed to be in the Perrier skull, and as would be expected in a primitive member of its tribe (elevated occiputs are a feature of highly derived sabertooths, as we shall see in the next chapter). This was a problem, because it either implied an evolutionary reversal in the putative descendant of *Megantereon*, *Smilodon fatalis*, which had an inclined occiput, or it meant a precocious specialization in the Senèze cat, which would then have to be excluded from the direct line of ancestry of *Smilodon*. This vexing problem was solved when Werdelin and I came across a cast of the Senèze skull in an unlikely place—the National Museums of Kenya, in Nairobi. The cast revealed something that one could not tell from looking at the original fossil in its exhibit case: the whole posterior part of the skull had been restored, and it had a completely different texture from the anterior part; it was smooth and relatively featureless, even bearing the traces of the restorer's fingers here and there. It is likely that the restorer was advised to follow *Smilodon populator* as a model for the damaged parts of the skull of *Megantereon*, since a cast of a complete skeleton from Argentina is conveniently housed at the Basel museum, and it has been actually exhibited alongside with

3.63. Reconstructed life appearance of *Megantereon whitei.*

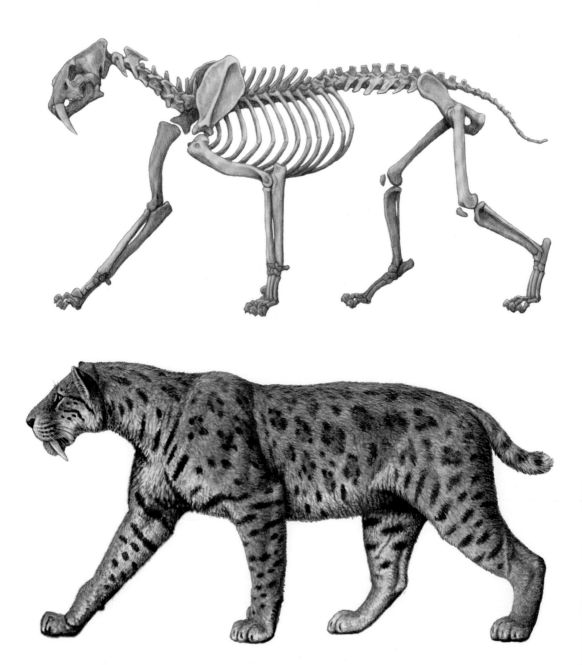

3.64. Skeleton (top) and reconstructed life appearance of *Smilodon fatalis*. Shoulder height: 100 cm.

Megantereon. A new skull of *Megantereon* from Dmanisi, Georgia, that was recently described (Vekua 1995) confirms the presence of an inclined occiput, as has also been proved by Chinese and African material (Antón and Werdelin 1998).

OTHER *MEGANTEREON* SPECIES

There is some disagreement about the specific classification of the Old World *Megantereon* from the Pliocene and Pleistocene. Although some specialists prefer to unify all samples in one single, variable species,

M. cultridens, others (Martínez-Navarro and Palmqvist 1995) see at least a second species, *M. whitei*, in African and some Mediterranean sites (figures 3.62 and 3.63). At least one other, allegedly primitive, species—*M. ekidoit*—has been described from the early Pliocene (3.5 Ma) site of South Turkwell, in Kenya (Werdelin and Lewis 2000). Chinese specimens have been attributed to *M. nihowanensis* and *M. inexpectatus*, but their differences with the European material are not clear-cut enough. The presence of the genus in North America is based on material of Blancan (early Pliocene) age attributed to *Megantereon hesperus*, which might have evolved in situ to give origin to *Smilodon* (Martin et al. 1988).

Genus *Smilodon*

As we saw in chapter 1, the genus name *Smilodon* was coined by P. Lund in 1842, on the basis of fossils from Lagoa Santa in Brazil, and over the years many species have been described. Most of those species names are now considered invalid, however, and most specialists recognize only an early species, *S. gracilis*, comparatively small and rather similar

to *Megantereon*; and two larger, later species, *S. fatalis* and *S. populator*. The genus is Pleistocene in age and exclusive to the Americas.

SMILODON GRACILIS

This is the earliest species of *Smilodon*. It was a jaguar-sized cat, not larger than the largest specimens of *Megantereon cultridens*, such as the one from Senèze. Best known thanks to a sample from the site of Haile in Florida (Berta 1987), the skeletal remains show that this animal was stocky and strong. Its dentition was more advanced than that of *Megantereon*, as were some aspects of its skull, including the mastoid region.

SMILODON FATALIS

This is the popular sabertooth cat from Rancho la Brea and other sites of late Pleistocene age in North America. It was a very large smilodontin, very similar to a lion in linear dimensions but with very robust, muscular limbs and body, implying a body mass considerably larger than that of a lion or tiger (figure 3.64). Like *Megantereon* it had a long and strong neck, a short back, and a stubby tail. The skull was broadly similar to that of *Megantereon*, but more massive, with larger upper canines, although it had a very reduced mental process in its mandible (figure 3.65 and 3.66).

The anatomy of *Smilodon fatalis* was described in admirable detail by J. Merriam and C. Stock in 1932, thanks to the spectacular sample of fossils from Rancho la Brea. As mentioned above, *Smilodon fatalis* was similar to a lion in its linear dimensions, implying that it would have a comparable total body length and shoulder height, but since it was more robust and muscular, it would have weighed considerably more. Precisely how much more is a difficult question to answer, and like modern big cats, the dimorphic *S. fatalis* would have been extremely variable in body weight. The paleontologist W. Anyonge (1993) estimated the body weight of *S. fatalis* on the basis of long bone dimensions, which are more reliable than the dental measurements traditionally used for mass estimates in fossils. His results indicate weights of between 340 and 440 kilograms, quite impressive if compared to a range of 110 to 225 kilograms for extant African lions. However, a more recent estimate by P. Christiansen and J. Harris (2005), based on thirty-six osteological variables, gives a range of 160–280 kilograms, still imposing but more in line with the weights of modern big cats.

The species name *Smilodon fatalis* was coined by Leidy, who in 1868 described a fragmentary maxilla form Hardin County, Texas, but, as some readers may already be guessing, it was not originally classified in the genus *Smilodon*—things are never that simple! Actually, Leidy thought his fossils belonged to a member of the genus *Felis*, although he considered them distinctive enough to put them in a new subgenus and species: *Felis (Trucifelis) fatalis*. After the customary taxonomic comings

3.67. Reconstructed life appearance of *Smilodon populator,* galloping. This huge sabertooth cat would use the walking or trotting gaits in most situations, but it would also be capable of short bursts of speed. In the gallop, the powerful muscles of the back are recruited for the run, implying an enormous expenditure of energy. With the largest individuals weighting nearly 400 kg., such an effort could be expected only when hunting, fighting for territory, or in pursuit of a potential mate. Shoulder height: 120 cm.

and goings, which have seen the rise and fall of species names like *S. mercerii,* *S. floridanus,* and *S. californicus,* it is now generally accepted that there were only two species of *Smilodon* in North America: the lion-sized, late Pleistocene *S. fatalis* and the smaller and older (from the early and middle Pleistocene) *S. gracilis,* intermediate in size and morphology between *Megantereon* and *S. fatalis.*

SMILODON POPULATOR

This species includes the largest and most robust specimens of *Smilodon,* and several of the larger individuals must have weighed more than 400 kilograms (figures 3.67 and 3.68). Apart from absolute size and robustness, the animal's differences with the North American *S. fatalis* are rather subtle, including a more straight dorsal profile in the skull, a more verticalized occipital plane (figure 3.69), and relatively shorter metapodials.

An amazingly complete skeleton from the late Pleistocene of the Buenos Aires region in Argentina was described in detail by R. Méndez-Alzola in 1941, a study that established the body proportions of this species and its striking differences with modern big cats—especially the tiger, which served the author as the standard for anatomical comparison.

This species was described by Lund in 1842 based on fossils from the late Pleistocene caves of Lagoa Santa in Brazil. Argentinian specimens have been named *Smilodon neogaeus* and *Smilodon necator,* but none of those names now appears to be valid. *Smilodon populator* occupied a huge range in South America east of the Andes, from Venezuela in the north to Patagonia in the south.

3.68. Reconstructed life appearance of *Smilodon populator.*

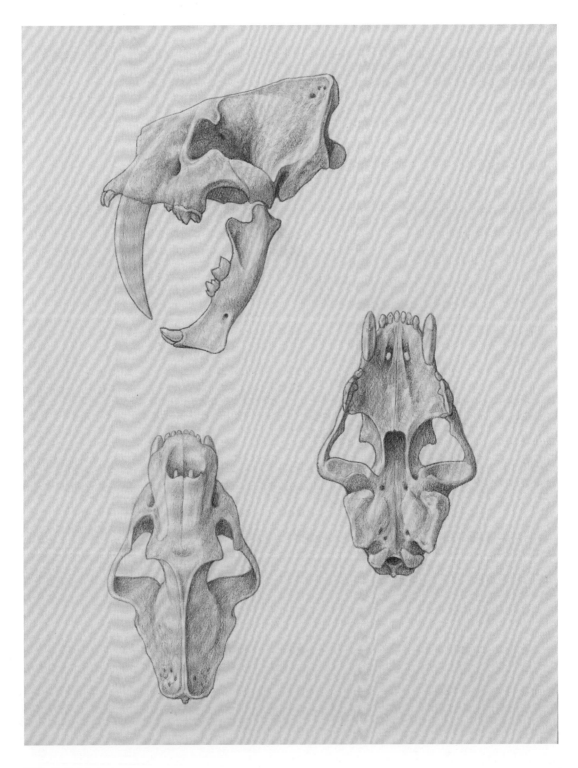

3.69. Skull of *Smilodon populator* in lateral (top), ventral (middle), and dorsal (bottom) views.

Sabertooths as Living Predators

4

ALL THAT HAS SURVIVED OF THE SABERTOOTHS ARE THEIR FOSSIL-
ized bones, but they once were living creatures, and the aim of the
science of paleobiology is to infer from their fossils as much as possible
about their ways of life. But can we really do more than just imagine how
the sabertooths moved, hunted, and interacted? Actually, if we know how
to look, fossil bones can yield a surprising amount of information. Using a variety of methodological tools including functional morphology,
comparative anatomy, dissection, and three-dimensional imaging, it is
possible to get a remarkably rich picture of the once living creatures. The
process is complex, and just like forensic scientists using the available
evidence to solve a crime, we have to seek a balance between intuition
and common sense.

The first step in this process is to reconstruct the anatomy of the
sabertooth from the inside out, starting with its skeleton, posture, and
proportions, and continuing with the musculature and the rest of the soft
tissues, including the skin and even coat patterns. After that, the next step
is to set the reconstructed creature in motion, inferring from the physical
traits of its locomotor system the likely gaits and athletic abilities of each
different sabertooth species: running, climbing, wrestling down prey, and
so forth. Cranial structures associated with the brain and nerves provide
information about coordination and sensory development. Combining
all these data with the information about prey species and characteristics
of the paleoenvironments, we can build hypotheses about the sabertooths' hunting methods, which will be enriched by data about injuries
and trauma, often associated with hunting accidents. Data on development and sexual dimorphism can give clues about family life and social
structure, rounding out the picture of sabertooth lifestyles.

Reconstruction

Vertebrate paleontology has a lot to do with puzzle solving. Given the
nature of most fossil sites, as discussed in chapter 2, the majority of the
material available to paleontologists consists of separate, often broken,
bones. It is thus easy to imagine the wonder of paleontologists when a
complete, articulated skeleton comes to light. Such exceptional specimens instantly become the standards we can use in putting together the
pieces of fragmentary finds of the same or related species. In the case of
mammalian sabertooths, we probably have reasonably complete skeletons
for fewer than twenty out of more than fifty recorded species.

Due to the scarcity of complete individuals, the work of reconstruction often starts by completing the missing parts of a skeleton. This task involves the reconstruction of the morphology of unknown parts and the scaling of pieces that belong to individuals of different size. Many sabertooth species varied considerably in individual size, which has to be accounted for during reconstruction. The relative lengths of the long bones of the limbs are essential in order to reconstruct body proportions, but there is an added problem: the longer the bone, the greater its chances of being broken before fossilizing. Museum collections abound in complete bones of the ankle or wrist of carnivores – small, squarish objects that survive well the processes of fossilization, but most of the long bones are broken. In these cases it is necessary to calculate ratios of length to width based on the proportions of comparable complete bones, to estimate how much of the bone's original length is missing.

For some species, genera, and even families, we lack relevant parts of their skeletons altogether. The history of sabertooths has plenty of such frustrating blanks. The American creodont sabertooth *Apataelurus*, the most derived genus in its family, is known from a single jaw, and in more than a century of excavations after its initial discovery, the extensive Eocene deposits of the Uintan still refuse to yield a single additional bone. Similarly, *Eusmilus sicarius*, the most spectacular nimravid sabertooth from North America, is known from cranial and mandibular remains only. These are extreme cases, but even when the blanks in a fossil animal's anatomy are comparatively small, we need solid criteria to reconstruct the missing parts. Besides knowledge of vertebrate anatomy, in many cases we need to resort to phylogenetic information, as will become clear below.

Phylogeny

Whether we are filling in some missing vertebrae of a sabertooth skeleton or deciding about a coat pattern for a life restoration, we are reconstructing unpreserved attributes of the fossils, and our first step is to refer to other taxa, as closely related as possible to the fossil, as models for the restoration. This principle informed, in an intuitive way, the work of early paleontological artists like C. Knight, but recent developments in phylogenetics have allowed us to refine the methodology. The theoretical basis for this procedure has been presented independently by different paleontologists (Bryant and Russell 1992; Witmer 1995), whose approaches differ in details but agree on two essential postulates: the reconstruction of unpreserved attributes in fossil taxa must consider the condition observed in the closest relative for which the attribute is known, and the inference becomes more robust if we can confirm the same condition in the next closest reference group, which is called the outgroup. Witmer coined the term Extant Phylogenetic Bracket for his methodology and proposed it as an approach for the restoration of all kinds of unpreserved attributes, but especially soft tissue, in fossil vertebrates. The condition

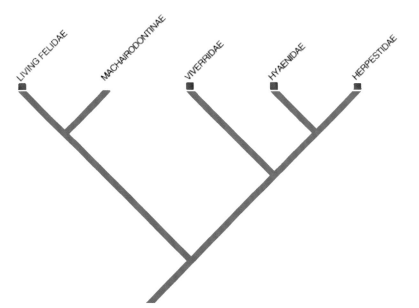

4.1. Cladogram showing the relationships of the felid sabertooth subfamily, the Machairodontinae, to its living relatives. When using the Extant Phylogenetic Bracket methodology, the condition of the unpreserved attributes observed in the closest group (in this case, the extant Felidae) is the most important reference, but if the same condition is observed in the outgroup (in this case, the clade formed by the other feliform families) then the inference becomes more robust.

of the unpreserved attribute in the reference groups provides the most conservative hypothesis about its condition in the fossil. The condition observed in the sister group is considered a good guide, but if it is similar in the outgroup as well, then our inference becomes considerably more robust (figure 4.1).

The second step is to look for any indications in the morphology of the fossil that may suggest the presence of a unique derived condition (called an apomorphy) of the unpreserved attribute in the fossil. Normally, the phylogenetic hypothesis takes priority (meaning that we assume the condition is the same as in the animal's close relatives), but if the evidence for a unique derived condition is compelling or if the phylogenetic evidence is ambiguous, then we proceed to make extrapolatory analyses, such as studies of form-function correlations. In the case of sabertooths, the differences with non-sabertoothed relatives generally don't imply the presence of radical differences in soft-tissue structures, but simply a rearrangement of the familiar structures to fit into a modified biomechanical context. There may be exceptions to this rule, however. In the case of barbourofelid sabertooths, the paleontologists V. Naples and L. Martin (2000) have proposed the presence of a derived condition in the arrangement of masticatory muscles, unseen in any known carnivore. According to these authors, the enlarged infraorbital foramen of these sabertooths provided room for the deep part of the masseter muscle, whose attachment in the skull (which in most mammals is placed on the cheek area of the malar bone) would have migrated to the area in front of the orbits, and whose fibers would pass through the foramen on their way there (figures 4.2. and 4.3). This is a condition observed in some rodents such as the porcupines of the genus *Hystrix,* and as a result it is known as hystricomorphy. Following this interpretation, such an unusual arrangement

4.2. Skull of *Barbourofelis fricki* with the deep masseter muscle reconstructed according to the "hystricomorph" hypothesis, with the fibers running through the infraorbital foramen.

of the masseter in barbourofelid sabertooths would be an adaptation to improve the mechanical advantage of this muscle when contracting from extremely wide gapes (which, it should be noted, is not exactly the same purpose that this adaptation serves in hystricomorph rodents). As we shall see below, there are other possible explanations for the presence of very large infraorbital foramina in sabertooths, but the possibility of a histrycomorph masseter in *Barbourofelis* remains a potential explanation, and it would make an interesting example of the detection of an unpreserved apomorphy in soft tissue through osteological form-function correlation.

Assembling Skeletons

The skeletons of carnivores can assume a wide range of postures, but when we draw a skeletal reconstruction we usually choose a "neutral" pose. Standing or walking postures are ideal for this type of "standard" skeletal drawings, but they require some important inferences about the way the bones articulated with each other. Fortunately, the skeleton of

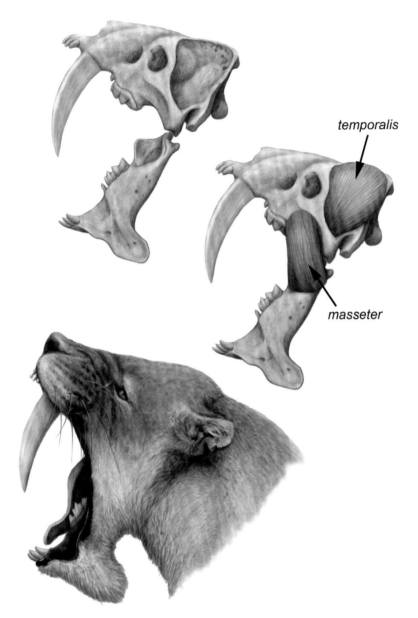

temporalis

masseter

terrestrial mammals is built as a tight system of levers, and in many bones the shape of the articular surfaces gives a fairly clear indication of the ranges of flexion and extension.

Thus, looking at the shape and orientation of the articular ends of vertebrae, it is possible to know if the neck had a strong "S" curve or was straight, or if the back was normally carried with a marked arching—although it is necessary to take into account the influence of the cartilaginous intervertebral discs, which in life occupy the spaces between vertebrae and modify the curvature of the column (figure 4.4). The articulations of the elbow and knee tell us if the animal stood and walked with flexed, crouched limbs, or if it used a more erect gait. The shape of

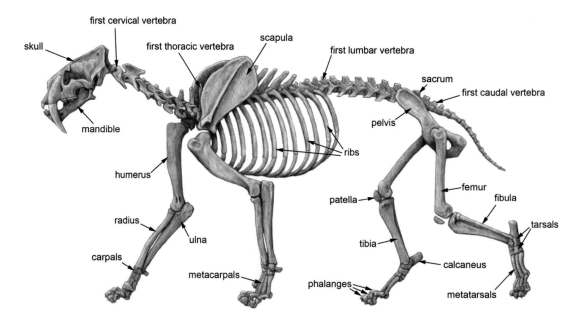

first cervical vertebra

skull

first thoracic vertebra

scapula

first lumbar vertebra

sacrum

first caudal vertebra

mandible

pelvis

ribs

humerus

femur

fibula

patella

radius

tarsals

ulna

tibia

carpals

calcaneus

metacarpals

phalanges

metatarsals

4.4. Skeleton of *Smilodon* with the main elements labeled.

the olecranon process in the proximal ulna is also telling in this respect, because it houses the tendon of the triceps brachii muscle, a major flexor of the arm. If the usual posture of the elbow is flexed (as in crouching extant carnivores such as palm civets and binturongs), the olecranon is bent forward; if the usual posture is extended (as in extant carnivores like the lion, cheetah, dogs, and hyenas), the olecranon is bent backward. Among mammalian sabertooths we find a wide range of variation, with the forward bent olecranon of the creodont *Machaeroides* suggesting a crouching animal that was an able climber, and the backward bent olecranon of the felid *Homotherium* pointing to a fully terrestrial animal that walked and ran efficiently on straight limbs (figure 4.5).

More difficult is to establish whether the feet of fossil carnivores were plantigrade or digitigrade. Among mammals in general, a digitigrade posture is usually associated with greater running abilities, because it makes the limbs functionally longer. Each stride the animal takes also becomes longer, and the area of the feet that contacts the ground becomes smaller, making each step more efficient. On the other hand, plantigrade locomotion is more stable and is often associated with animals that need strength rather than speed, as is the case with those that dig (like bears or badgers among carnivores) or climb habitually (like coatis or palm civets).

There is a clear difference between the appearance of a plantigrade carnivore like a bear, which like humans appears to rest the entire sole of the hind feet on the ground while walking, and the fully digitigrade stance of a dog, with its heels well clear of the ground and its foot bones oriented vertically (figure 4.6). Not surprisingly, the foot bones of dogs and bears are dramatically different. But on detailed observation, it is obvious that the situation is more complex: in "pure" plantigrades such as humans, the heel bone or calcaneum touches the ground first when stepping, and the sole of the foot has a concave, bridge-like shape; but the tip of the bear's calcaneum is always slightly raised above the plane of the other foot bones when they are planted on the ground.

To complicate matters, there are several living carnivores, including some viverrids (members of the civet and genet family) and some mustelids (members of the weasel family), that are not really plantigrade because they don't rest the heel on the ground when walking, but their feet bones are not vertical when touching the ground and can even be close to horizontal. It can be tricky to determine with the naked eye the stance of some of these carnivores, and it may require a close examination of filmed footage and a study of footprints to confirm the position of the feet. In fact, although these animals are alive and kicking around us, scientists have been confused about their gaits, so for a long time it was claimed in scientific papers that genets are plantigrade animals, which they are not, although they use a more nearly plantigrade stance when moving along the branches (for increased stability) than when they walk on the ground. In these animals, which we could call "low-angle digitigrades," the shape of foot bones is also intermediate between that of the most typical plantigrades and digitigrades. Many sabertooths had precisely such intermediate morphologies in their foot bones, so it comes as no surprise that determining their posture is a tricky business.

One interesting example is the Pleistocene scimitar-tooth *Homotherium*, which for years was reconstructed with plantigrade hind feet. The hind limb bones of *Homotherium* do have some morphological features comparable to those of ursids (such as a short calcaneum and a relatively flat astragalus), which led some specialists to think that the animal would walk on bear-like, plantigrade hind feet. But a detailed study conducted by the French paleontologist R. Ballesio (1963) on the complete skeleton of *H. latidens* from Senèze made it clear that there was an overwhelming number of morphological traits indicating a digitigrade hind foot in *Homotherium*. These included the shape and arrangement of the metatarsals, which are long, straight, and parallel, with the two central ones noticeably longer than the others–in contrast to the more fan-like arrangement of metatarsals in typical plantigrade carnivores.

How, then, can we explain the apparent plantigrade features of *Homotherium*? They are probably related to the animal's need for increased stability during the hunt (as we shall see below) rather than to its posture during normal locomotion. Defining the foot posture of a sabertooth may sound like a subtle matter, but the implications for locomotion and behavior are important, and the differences in the resulting appearance of the reconstructed animal can be striking (figure 4.7).

In spite of past confusion, the case of *Homotherium* is now rather clear, because there are so many features indicating a digitigrade posture. But in other cases it is harder to decide. *Xenosmilus*, a close relative of *Homotherium*, had very robust limbs with short, broad feet (Naples 2011). The overall morphology of these feet has many traits in common with the feet of plantigrade carnivores like bears, and it is difficult to decide if

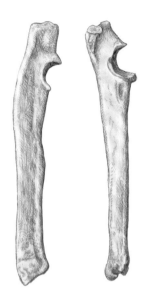

4.5. Comparison between the shape of the proximal ulna in a climbing sabertooth, the creodont *Machaeroides eothen* (left), and a highly terrestrial species, the felid *Homotherium latidens* (right). Note that the caudal border of the olecranon is convex in the first case, and concave in the second. The bones are not drawn to scale.

4.6. Comparison of a plantigrade carnivore, the raccoon (top), and a digitigrade carnivore, the domestic cat (bottom). The hind limb in each animal is shown in transparency to reveal the disposition of the bones inside. The arrows show the position of the calcaneum.

Xenosmilus walked on flat feet or not. The posture of members of completely extinct families like the barbourofelids and nimravids can also be tricky to reconstruct. Studying the foot bones of the barbourofelid *Sansanosmilus*, from Sansan, the French paleontologist L. Ginsburg (1961b) concluded that it would have been a plantigrade animal, but he believed that the true felid *Pseudaelurus*, from the same Miocene site, would have been digitigrade like modern felids. Ginsburg's conclusions were not easily reached: he conducted a thorough study of modern carnivores and established a series of traits distinguishing the feet skeletons of plantigrade carnivores from those of digitigrade carnivores. Ginsburg's criteria remain hallmark references for the interpretation of foot morphology in carnivores, but there is an important difficulty with morphological criteria: plantigrady in carnivores is usually associated with slow progression and digitigrady with running (cursorial) adaptations; morphology is more likely to tell us if an animal was more or less adapted to cursoriality (running) than to help us place it on one side or the other of the plantigrade-digitigrade divide. Some animals are not especially fast and yet are perfectly digitigrade, and those animals are especially difficult to classify as plantigrades or digitigrades on the basis of their morphology alone.

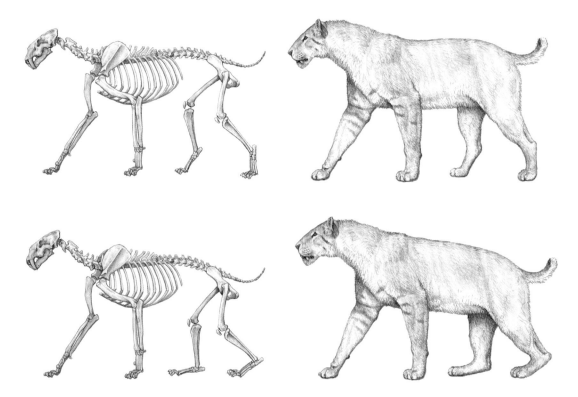

It is obvious that the majority of nimravid and barbourofelid saber-tooths, and even the marsupial *Thylacosmilus*, were not highly digitigrade animals with vertical feet like modern cats. However, their exact foot posture is a matter that clearly requires further study.

4.7. Alternative reconstructions of the skeleton (left) and life appearance (right) of *Homotherium latidens* to show the digitigrade (top) and plantigrade (bottom) hypotheses. Note that in the plantigrade version, the stride of the hind legs is so short that they would almost have to jump in order to keep up with the longer stride of the forelimbs during the walk.

A Note on Gait and Fossil Footprints

Although fossil footprints attributed to sabertooths are very rare, the paleoichnological record (meaning the record related to fossil footprints) still has something to add to our picture of sabertooth locomotion. Isolated footprints that may have been made by felid sabertooths of the genera *Nimravides* and *Machairodus* have been found in at least three sites of late Miocene age in California: the Death Valley National Monument, the Avawatz Mountains, and the Mojave Desert (Alf 1959, 1966; Scrivner and Bottjer 1986). These footprints average 9–10 centimeters in total length and thus correspond to animals about the size of a small lion. Their shape is catlike, with paraxonic disposition of the digits (meaning that the two central digits are of similar length, and relatively longer than the lateral ones), a semicircular to subtriangular main pad, oval digit pads, and no claw marks, which indicates the presence of retractable claws. Since no conical-toothed cats had attained a size larger than that of a lynx in the Miocene (and there are footprints of lynx- and wildcat-sized felids in the Death Valley deposits as well), these footprints can only be those of sabertooth cats. Their very modern, digitigrade morphology argues against attributing them to the barbourofelids, which also inhabited North America in the Miocene, but whose foot morphology indicates at least a semi-plantigrade posture, with a larger foot surface resting on the ground during the walk.

Perhaps the best preserved footprints attributed to a sabertooth cat come from the Hemphillian (late Miocene) site of Coffee Ranch, in Texas. Fossil remains of *Amphimachairodus coloradensis* have been found at the site as well, making it very probable that the tracks belong to this species. These footprints are very similar to those of a modern cat in morphology, and with a total length of about thirteen centimeters, they are larger than the Californian tracks mentioned above, suggesting an animal the size of a very large lion or tiger (Johnston 1937).

Older footprints found at an early Miocene site in Spain point to an earlier stage in felid evolution. The site, called Salinas de Añana and located in Alava (Basque country), preserves in exquisite detail an exceptional sample of carnivore tracks, including those of two clearly catlike species (Antón et al. 2004a). The larger of these species, about the size of a lynx, had feet essentially similar to those of modern cats, but with a relatively larger main pad, which indicates a less purely digitigrade posture. The long series of tracks found at Salinas de Añana give us our first clear glimpse into early Miocene felid locomotion, which included lateral-sequence walks and diagonal-sequence trots like those observed in the animals' modern relatives. A set of parallel tracks found at the site indicates group travel, probably related to the presence of family groups composed of an adult female with adult-sized cubs. These tracks give us a rare look at early cat behavior, and they suggest that the modern pattern, in which cubs remain associated with their mother until reaching adult size, was already established 15–20 Ma.

Fossil footprints are classified according to a parataxonomy – that is, a classification scheme independent from that of living animals or their direct remains. Such a procedure is useful for ichnologists to agree about what type of footprints or trace marks they are dealing with, independently of the difficulties of attributing them to concrete fossil taxa. This parataxonomy is purely morphological, so if two footprints are indistinguishable in shape they will be attributed to the same ichnotaxon even if there are reasons to believe, on geographical or stratigraphical grounds, that they were produced by different species of animal. Thus, although some of the American footprints have been informally referred to as "cat footprints," they are formally classified in the ichnogenus *Felipeda*, just like the older and smaller tracks from Salinas de Añana. It is difficult to know if the cats that left the Salinas tracks were ancestral to sabertooths or if they were members of the feline branch of the felidae, but the Salinas tracks clearly indicate the presence of a lynx-sized cat with a somewhat primitive foot posture that nonetheless had a locomotion pattern basically identical to that of modern cats. The American footprints, in contrast, confirm that felids of late Miocene age had developed modern foot morphology and a completely digitigrade stance, which corresponds to data from fossils

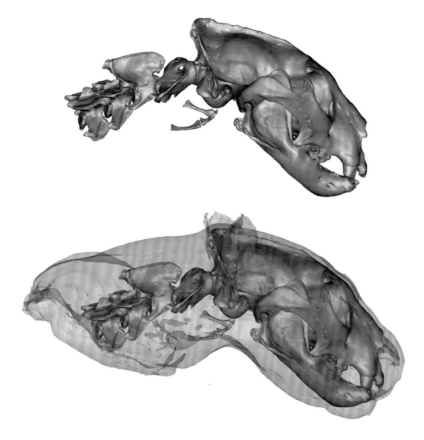

4.8. CT scan images of the head of a modern leopard (*Panthera pardus*), revealing the angles of articulation between the skull and cervical vertebrae (top), as well as the relationships between bone and soft tissue (bottom).

of sabertooth felids such as *Nimravides* and *Amphimachairodus* from the same age and place.

Reconstructing Soft Tissue

Once we have assembled a skeleton in a lifelike pose, we can start adding the soft tissues. First, we reconstruct the deeper layers of muscle, using as a guide the marks, ridges, and roughened areas on the bone that indicate the presence of muscle insertions.

The main muscle masses, which determine the broad outline of the living animal, can be derived from the skeleton in a relatively straight-forward manner. Of course the dissection of extant carnivores provides the main guide for this labor, but dissection is not without limitations. It is a destructive process, so that once we have taken away the superficial muscles we cannot put them back in place, although naturally the only way to see what lies beneath a layer of tissue is to peel it off. These limitations are problematic when we want not only to observe the hidden details of an animal's anatomy but also to check the spatial relationships between the skeleton and the external features. One useful technology to help us avoid this problem is CT (meaning computed tomography) scanning, which allows us to observe at the same time and in three dimensions both the soft tissues and the bones inside them, and without

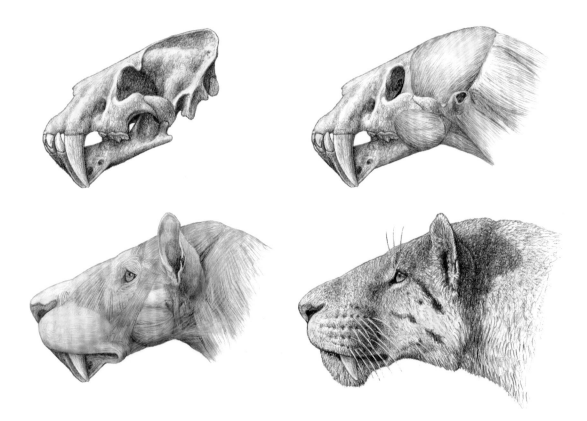

4.9. Sequential reconstruction of the head of *Homotherium latidens*. Top left, skull and mandible; top right, deep muscles; bottom left, superficial musculature and cartilages; bottom right, external appearance. Note that, in addition to the tips of the sabers that protrude beyond the ventral margin of the upper lip, the appearance of the head differs from that of any modern big cat because of its straight dorsal outline and its large muzzle.

the need to destroy the specimen. This technique is especially useful in the reconstruction of the heads of sabertooths, but it is also helpful for the whole post-cranial anatomy (figure 4.8).

Starting with the head, we find that in the skulls of sabertooths the big masticatory muscles, the masseter and temporalis, fit in well-defined areas bordered by the sagittal and lambdoid crests, the masseteric fossa, and other salient features (Antón et al. 1998, 2009). This is fortunate because the head embodies so much of the "personality" of any animal species, and the main muscle masses go a long way in determining the volumes of the living animal's head (figure 4.9).

With the shape and position of the main head muscles defined, we continue with the more superficial muscles as well as non-muscular features that are essential to the appearance of the living carnivore, including the lips, skin, and cartilages of the nose and ears. These tissues are more complicated to infer from osteological morphology, and a higher level of inference (that is, more extrapolation) is necessary. During the process of reconstruction, it is useful to create informal sketches that use color codes to show the relationships between the osteology of the animal and the different types of soft tissue (figure 4.10).

Years ago, a peculiar interpretation of the relationship between cranial morphology and external features in *Smilodon* produced a rather odd picture of the living head of this animal—and, by extension, the heads of other sabertooths—which received some acceptance among

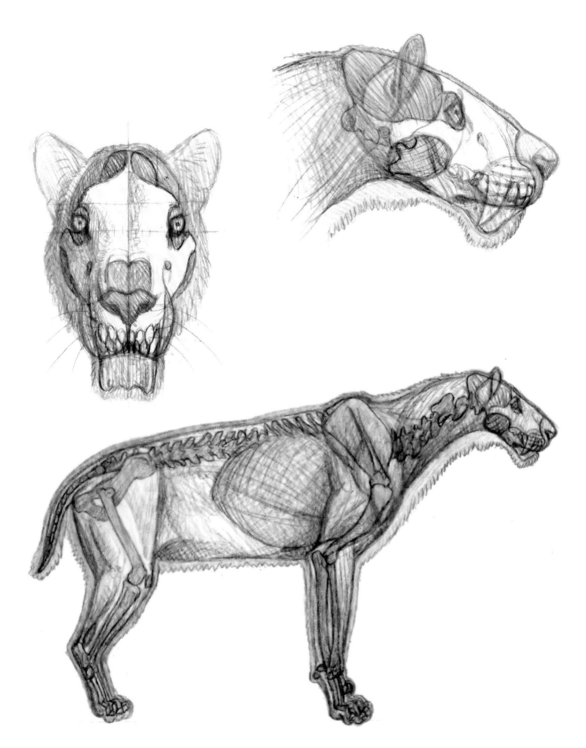

specialists and artists in the later part of the twentieth century. In 1969 the American paleontologist G. Miller challenged the catlike appearance of "traditional" reconstructions of *Smilodon*, such as those created by Knight in collaboration with J. Merriam and C. Stock (1932). Miller pointed to some differences between the skulls of *Smilodon* and those of modern big

4.10. Informal sketches of *Homotherium serum* head and body reconstruction, with color codes showing the bone in blue; the muscles, fat, and cartilage in reddish brown; and the skin and fur in green.

4.11. Reconstructed life appearance of the head of *Smilodon fatalis,* according to Miller's hypothesis.

cats, including the retracted nasal bones (relative to the anterior end of the premaxilla) and high sagittal crest in *Smilodon,* which in his opinion implied that the living animal would have had a shortened, bulldog-like external nose, and that the external ears would look strikingly different because of the low position of the auditory meatus relative to the dorsal outline of the skull. Miller further proposed that the sabers would have gotten in the way of food items if the animal tried to bite with the side of its mouth, and he hypothesized that in order to overcome this difficulty, *Smilodon* would have evolved a long lip line, reaching much farther back than that of modern cats – an adaptation that, in Miller's view, would also have allowed the animal to achieve the large gapes necessary in order to bite with the sabers. Miller instructed an artist to produce a drawing of the face of *Smilodon* that reflected his hypothesis, and it certainly was quite different from Knight's rendering (figures 4.11 and 4.12).

The points raised by Miller pose special difficulties because the nose, ears, and lips leave little to no recognizable marks on the bone. Is it possible to decide between Miller's interpretation and that of Merriam and Stock, and Knight? To solve this riddle, we need to combine anatomy with the phylogenetic methodology outlined above. To that end, we first consider the relationships between skull morphology and external features in *Smilodon*'s living relatives (Antón et al. 1998).

Regarding the external, cartilaginous nose, Miller's proposal implies that the nose's anterior projection would be directly proportional to the position of the anterior margin of the nasal bones, but a review of modern felids indicates that this is not the case. Rather, the position of the rhinarium (the external opening of the nose) is linked to that of the premaxilla and incisor battery, and in spite of wide variations across species in the relative retraction of the anterior margin of the nasals, in all cases the cartilage spans the distance. To give an example, the difference in nasal bone retraction between the modern tiger and the lion is greater than the difference between the lion and *Smilodon*, but in both modern cats the rhinarium is similarly positioned on top and slightly ahead of the incisors, and the external appearance remains similar. Actually, Miller's choice of the "pug nosed" bulldog as a modern analogue was rather unfortunate, because this is a domestic breed in which artificial selection has created a pathological prognathism and an exaggerated retraction of the face and loss of proper occlusion between the upper incisors and the lowers, a condition not seen in any healthy wild carnivore. This example also demonstrates the importance of choosing analogues within a restricted phylogenetic group, because in other mammal groups the conditions may be different from those observed in carnivores. Great apes, for example, display retracted nasals that are associated with short external noses. This may be related to deep functional differences such as the reduced importance of olfaction in apes compared to carnivores, but the causes for the differences remain largely unknown. The only way to make sure that we are transferring the proper condition to fossils is to base our inferences in the correct phylogenetic context.

The external ear, or pinna, in modern cats emerges a short distance above the skull's external auditory meatus, so it is relatively simple to check if the position of the ears is right in a reconstruction. Miller had been right to point to the low position of the ears (relative to the high dorsal outline of the skull) in *Smilodon*, but the fact is that Knight's drawings showed them that way, although their position did not give his restorations the odd appearance evident in Miller's reconstruction. Actually, hyenas and other modern carnivores also have ears that emerge low relative to their very high sagittal crests, and they still look like "normal" carnivores. The reason why the ears in Miller's *Smilodon* looked so odd was not because of their low position, but because their morphology was simply wrong. The ears in modern cats—as well as in their relatives, the genets and civets—are morphologically very consistent, with the only major variation being relative size. Furthermore, they conform to a widespread pattern that is probably primitive for all carnivores, including a bag-shaped structure in the back margin called the bursa. Transferring such widespread ear morphology to sabertooths is a robust inference.

In modern cats, the posterior border of the lip line while in a relaxed position is slightly anterior to the anterior fibers of the masseter muscle. Considering that living felids are the extant sister group to the extinct subfamily Machairodontinae, it seems safe on phylogenetic grounds to transfer the observed relationships between bone and soft tissue to the fossil species, but that inference would be even more robust if the same relationships existed in the outgroup. We can choose as our outgroup the next closest relative of machairodontines, which might be either the viverrids or the hyenids. The choice would make no difference in this

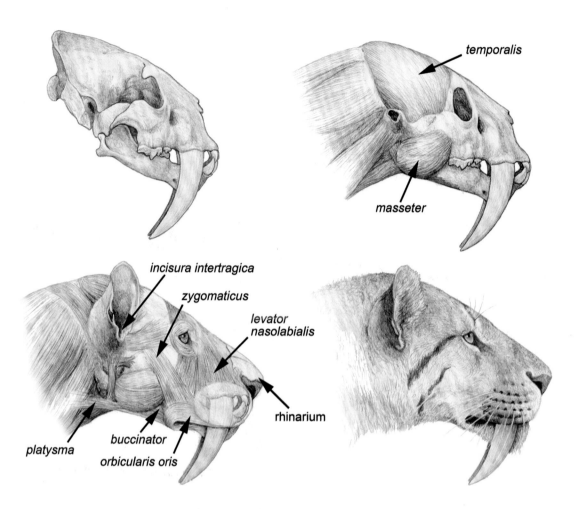

incisura intertragica
zygomaticus
levator nasolabialis
temporalis
masseter
rhinarium
platysma
buccinator
orbicularis oris

case, because in both families the examined features are the same as in cats–in fact, the same condition is observed in all modern members of the order Carnivora.

If the aperture of the mouth in *Smilodon* had been more extended posterior to the anterior margin of the masseter, it would still not have provided the animal with the benefit hypothesized by Miller–that is, being better able to bite pieces of meat with its carnassials. After all, the masseter muscle itself would then get in the way of any food item! Furthermore, simple observation of modern cats shows that their canines, although smaller than those of sabertooths, get in the way of food items just as would be the case with *Smilodon*, because in the carnassial bite, the gape of the jaws is so small that there is no clearance between the canine tips. The problems for food ingestion that Miller thought the sabers would cause to *Smilodon* were simply imaginary (figure 4.13).

Furthermore, the idea that a longer mouth aperture was necessary to provide for large gapes revealed a limited observation of modern animals. Hippos and, to a lesser degree, peccaries open their mouths to enormous gapes (well beyond 100 degrees, in the case of hippos) in order to display their canines, but they have "normal" mouth apertures that never reach behind the anterior margin of the masseter – the tissues of the mouth walls are simply more flexible than we tend to imagine. Even lions or tigers, when yawning with their jaws at gapes of around 70 degrees (a point where one would think that their mouth walls are as stretched as can be), can pull their lips back to bare their teeth in a grimacing gesture, showing that their stretching ability is far from having reached its limit.

Modern anatomical and phylogenetic approaches thus vindicate the careful work of Knight and his advisors back in the 1930s and confirm the catlike appearance of *Smilodon* apparent in his illustrations.

One interesting example of the application of the methodology just outlined is the possibility it provides of checking the identity of an animal depicted in prehistoric art. Such is the case of a Paleolithic statuette discovered in Isturitz, in the French Pyrenees, which – according to the Czech paleontologist V. Mazak (1970) – could be a portrait of the sabertooth *Homotherium*. One interesting implication of this attribution was that, since the sculpture shows no trace of the upper canine tips, Mazak inferred that the canines of *Homotherium* would be covered in life by the lower lips, unlike the condition that we see in modern cats or any other carnivore. A careful reconstruction of the soft anatomy of the head of *Homotherium* following the principles of comparative anatomy and the Extant Phylogenetic Bracket (figure 4.9) reveals important differences between the life appearance of the sabertooth head and the sculpture, strongly suggesting that the animal represented was, in fact, a lion (Antón et al. 2009).

If we turn to the whole body reconstruction (figure 4.14), we find that the main flexors and extensors of the legs and back are the largest and heaviest muscles of the cat's body, and they define the main ways in which the outline of the living animal is different from that of its skeleton.

Many of the more superficial muscles leave little or no mark on the bones, and their position is best inferred from the condition seen in comparable modern species and relative positions of the insertions of other muscles (Barone 2010). We can thus develop a reasonable image of the whole musculature (figure 4.14).

Coat Color Patterns

When trying to reconstruct the unpreserved coat patterns of extinct mammals, phylogenetic evidences have to be considered first. In the case of carnivores, certain patterns are widespread among some groups and rare in others. Thus, spotted or striped body patterns are virtually absent in members of the caniformia (the carnivore suborder including dogs, bears, and weasels), but fairly widespread among the feliformia (cats, hyenas, civets, and mongooses). This difference is rather clear-cut and gives a broad indication of what patterns not to use when dealing with extinct members of each suborder (banded tails, however, are known in members of both groups). Then there are some functional considerations. For instance, dark facial masks and contrasting adjacent light-colored

4.12. Sequential reconstruction of the head of *Smilodon fatalis*. The position of the main muscles of mastication, the temporalis and masseter (top right) is clearly indicated by features in the skull (top left), and their mass goes a long way to define the volumes of the animal's living head. The positions of some of the more superficial muscles, like the levator nasolabioalis and the zygomaticus, are also tied to osteological features, and although they are thin and do not greatly modify the three-dimensional shape of the head, their trajectory is important when reconstructing facial expressions. Other superficial muscles like the platysma, buccinator, and orbicularis need to be reconstructed on the basis of the condition observed in modern relatives. The most ventral part of the ear opening, or incisura intertragica, is placed immediately above the external auditory meatus, defining the placement of the external ear, or pinna. The cartilaginous nose is placed so that its anterior tip, or rhinarium, is slightly anterior to the incisor arch (bottom left). External attributes like fur length and coloring are based on analogies with extant relatives, phylogenetic reasoning, and functional considerations (bottom right).

4.13. Skull with reconstructed outline of soft tissue (top) and reconstructed life appearance of the head of *Smilodon* (bottom), applying the carnassial bite to a carcass. As the drawing shows, the lateral position of the carnassials allowed the animal to bite directly with them at the carcass, without the canines' preventing the animal from acquiring the food. Also notice that the gape necessary for the carnassial bite is so small that even in modern big cats it is not enough for the canine tips to clear, so they are in the same situation as the sabertooth, and they don't need weirdly shaped lips to perform the bite – nor did *Smilodon*.

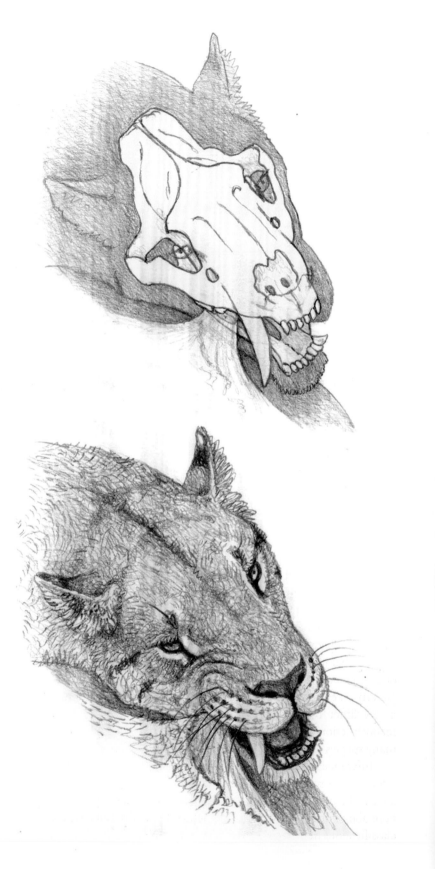

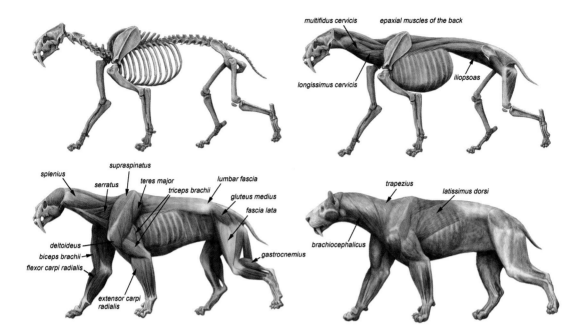

Labels in figure (top left): multifidus cervicis, epaxial muscles of the back, longissimus cervicis, iliopsoas

Labels (bottom left): splenius, supraspinatus, serratus, teres major, triceps brachii, lumbar fascia, gluteus medius, fascia lata, deltoideus, biceps brachii, flexor carpi radialis, extensor carpi radialis, gastrocnemius

Labels (bottom right): trapezius, latissimus dorsi, brachiocephalicus

areas are a common pattern among small to medium-sized carnivores with omnivorous diets and crepuscular or nocturnal habits, no matter what suborder they belong to (Newman et al. 2005). This includes such diverse modern carnivores as civets, raccoons, and badgers. Dark spots and stripes on a light background provide effective camouflage and are more widespread among forest-dwelling feliforms, but there are conspicuous exceptions and it is clear that although a spotted coat is a help for woodland-loving predators, some do perfectly well without one. Just as important is the fact that the spotted coats are widespread enough to suggest they were a primitive condition for cats, civets, and hyenas, an inference further supported by the fact that spots are present in the young of species (such as lions and cougars) whose adults have plain coats.

The implications of these observations about the coats of modern species for the reconstruction of sabertooths are various. For example, felid sabertooths probably shared an ancestral, genet-like pattern of spots and stripes with cats, their extant relatives, as well as with the genets and hyenas. Individual species may of course vary from that pattern, and it is likely that in some of the largest species, especially those living in the open, the spots would tend to fade or disappear. The evolution of unusual patterns like the mane of male lions and the vertical stripes of the tiger is very difficult to predict, so our rather conservative method implies that reconstructions are likely to reasonably approximate the appearance of many species . . . and to grossly fail with a few of them (figure 4.15).

Inferences about the coat patterns of the nimravids are particularly difficult because of their unclear relationship with other carnivores. If they are classified as feliforms, then we are justified in making similar assumptions about them as we do about the felid sabertooths. But if we accept the classifications that place them in a sister group to all the other carnivores,

4.14. Sequential reconstruction of the body of *Smilodon populator.* Top left: skeleton; top right: deep musculature, including the epaxial muscles (those closest to the vertebral column), the intercostalis, and some deep muscles of the proximal hind limb; bottom left: deep muscles, including the masticatory muscles of the head, the main muscles of the limbs, and the lumbar fascia; and bottom right: superficial muscles, including facial musculature of the head, brachiocephalicus, trapezius, and great dorsal, as well as fascias and cartilaginous structures such as the pinna, nasal cartilages, and whisker pad.

then a spotted feliform pattern is no more likely than a plainer dog-like or bear-like coat. So, especially with the nimravids and all other sabertooth predators that lack particularly close living relatives, one is left to use common sense, reasoning based on function, and a bit of imagination.

The Brain and the Senses

The brain of mammals is protected by a thick wall of cranial bones that are pressed tightly against it. The fit is so close that the inner cavity of the skull matches the shape of the brain and associated structures in considerable detail. Thus, it is possible to know the external shape of the brain in fossil mammals as long as the inner walls of the skull are well enough preserved, in which case it is possible to fill the cavity with some suitable material and thus create a cast of the encephalon. In some fossil skulls, the brain cavity is completely filled with matrix, which reproduces the shape of the brain and is called a natural endocast. Using digital three-dimensional imaging technologies, we can produce a CT scan of a well-preserved fossil skull and observe the shape of the endocranial cavity on the computer screen.

A review of brain shape in sabertooths and related forms reveals that the Oligocene nimravids, as well as the early Miocene true felid *Proailurus*, had a relatively simpler pattern of *sulci*, or furrows, and *gyri*, or convolutions, than we see in modern cats, implying less brain complexity (Radinsky 1969). Felid sabertooths from the late Miocene onward, like *Machairodus*, *Homotherium*, and *Smilodon*, already display sulcal patterns of the modern type, showing an increased complexity in the regions that control hearing, eyesight, and limb coordination.

There is some variation in the relative size of the brain among extant cats and also among machairodontine felids, and some early studies suggested that the brains of lions were larger in their linear measurements than those of closely related big cats such as tigers or jaguars (Hemmer 1978). This finding led to the suggestion that the larger brain of the lion could be related to the greater intelligence required by the animal's social lifestyle, while the smaller brain of the tiger would fit its solitary habits. Since *Smilodon* had a smaller brain relative to its body mass than *Homotherium* did, it was further suggested that the former was a solitary cat like the tiger, and the latter a sociable one like the lion. These assumptions should have been suspected to be oversimplifications to start with, because there is no proof that the tiger is significantly less smart than the lion, and because both animals have similarly complex brains in spite of any slight differences in relative size. In addition, differences in relative brain size between *Smilodon* and *Homotherium* are just as likely to reflect their different body builds (with the former being much more robust and muscular than the latter) as any measurable difference in intellectual prowess. But the strongest blow to the theory that brain size is correlated to sociality was delivered by a recent study (Yamaguchi et al. 2009) that measured brain volume (instead of linear measurements, as in previous studies) in a large sample of big cats and showed that, in fact, the brains of tigers are larger, relative to body size, than those of lions! Observations of the behavior of modern large felids in the wild certainly suggest that ecological constrains are probably a much more important determinant of social behavior in predators than any differences in the ratio of brain to body mass (Packer 1986; Packer et al. 1990; Sunquist and Sunquist 1989).

Since creodonts are often seen as the primitive forerunners of the true carnivorans, their brains could be expected to be simpler and smaller than those of early true carnivores, but that is not exactly the case. As

4.15. Alternative reconstructed life appearances of *Smilodon populator* with spotted (top) and plain (bottom) coats. Both alternatives are possible, and the choice between one or the other is often made on the basis of function. For instance, if the animal is supposed to have lived in forested habitats, the retention of a primitive spotted pattern is more likely. But there are examples to the contrary among extant cats, and coat color attributions remain no more than educated guesses.

happened in members of the order Carnivora, the neocortex of creodonts enlarged over geologic time, leading to the apparition of sulci, and although some differences are observed, these are not easy to interpret in biological terms. Relative brain size in Paleogene creodonts is generally similar to that in contemporary true carnivores.

As for marsupial sabertooths, endocranial casts of *Thylacosmilus* reveal a sulcal pattern broadly similar to that of modern marsupials (Goin and Pascual 1987).

Felid sabertooths differ from their modern relatives in the comparatively smaller size of their orbits. The big eyes of modern cats are associated with their crepuscular and nocturnal habits, and it is possible that ancestral sabertooths were more diurnal creatures. Modern cats also have frontally oriented orbits, which allow for a remarkable degree of binocular vision, an adaptation important in estimating distances during the hunt and while moving along the branches of trees. The eyes of machairodontines were slightly more laterally placed, although they still permitted a reasonable degree of binocular vision, comparable to that of a wolf. This is one of several features of felid sabertooths that suggests these animals had a longer history of terrestrial adaptation, compared to their more arboreal feline cousins.

One consistent feature of most sabertoothed carnivores is the relatively large size of the infraorbital foramen. As we have seen above, the extremely large openings in *Barbourofelis* have led to the hypothesis that fibers of the masseter muscle passed through that opening, in a pattern resembling the condition observed in some modern rodents. In *Barbourofelis* not only is the opening huge, but there is also a broad area in the maxilla in front of it that looks like a muscle insertion, which is where the hypothetical fibers would attach. However, most sabertooth carnivores have relatively large openings (if not as huge as in *Barbourofelis*), but without any indication of muscle attachments in front of them. Although the passage of muscle fibers through the infraorbital foramen in mammals is a derived condition found only in some rodents, there are other structures that pass through this opening in all mammals – specifically, the infraorbital nerve, veins, and arteries. The infraorbital nerve is a branch of the maxillar nerve, and it provides sensory nerve endings to the whiskers. Once the nerve crosses the canal, it begins to branch; and when it reaches the roots of the whiskers, it forms a true "nerve pad," creating a characteristic swelling on the sides of the muzzles of many mammals. Thus, one tempting explanation for the large diameter of the opening in sabertooths would be the presence of very well developed, especially sensitive whiskers with rich innervation. One interesting example among modern carnivores is the aquatic species, which usually have thick, extraordinarily sensitive whiskers that they use to sense variations of pressure in the water around them and to assess the position of prey in conditions of low visibility. Such species, including the pinnipeds but also the otters, have consistently large infraorbital foramina, and even the otter civet of Asia – a semiaquatic viverrid with huge vibrissae, or whiskers – has larger

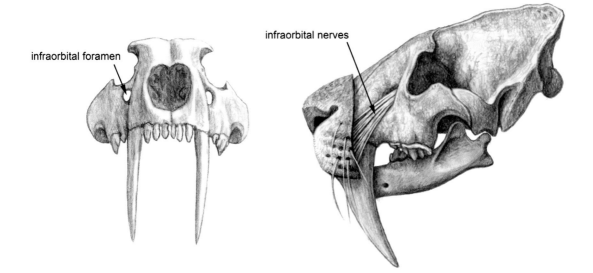

infraorbital foramen

infraorbital nerves

foramina than its non-aquatic relatives. In contrast, modern bears and hy-enas, which have independently evolved small, almost vestigial whiskers, display correspondingly narrow, small infraorbital foramina (figure 4.16).

Unfortunately, the relationship between the diameter of the infra-orbital foramen and the thickness of the infraorbital nerve is not one of simple, direct proportionality. The combined diameter of the nerve and blood vessels is smaller than the hole they pass through, and the rest is filled with connective tissue, so it is not possible to use the size of the foramen as a quick way to estimate the diameter of the nerve in the fossil species. Additionally, although broad differences in foramen diameter between different carnivore families are at least suggestive of a relation-ship, some differences between species are difficult to explain. Lions, for instance, have consistently larger openings than tigers, but their whiskers are not larger, and there is no evidence that these whiskers are more sensitive than those of tigers. In any case, the relationship between the development of the foramina and the function of the infraorbital nerve is widely accepted among zoologists. For instance, in the case of primates, the reduction of the foramina among advanced anthropoids is seen as a correlate of the atrophy of whiskers and decreased tactile innervations in the muzzle region, as compared with the more primitive prosimians (Muchlinski 2008).

We have every reason to reconstruct sabertooth carnivores with prom-inent whiskers like those of modern cats. But is it possible that the saber-tooths' whiskers were more sensitive than those of extant cats, and if so, what would be the functional cause? As we shall see below, the killing bite of sabertooths must have been quite precise in order to avoid accidents involving lateral torsion or hitting a bone in the prey, which could cause the sabertooth to break a canine. Since the target area of the bite would be outside the predator's visual field, the tactile information provided by the whiskers would be especially useful for the precise control of the biting

4.16. Left: frontal view of the skull of *Megantereon cultridens,* showing the infra-orbital foramen. Right: lateral view of the skull of *Smilodon fatalis* with the reconstructed infraorbital nerve exiting the infraorbital foramen; and the partly cut soft tissue of the muzzle, including the whisker pad and whiskers. As in mod-ern carnivores, we infer that the nerve has already started to branch before exiting the foramen, and that it continues to branch as it approaches the bases of the whiskers.

motions. Modern cats are able to move their whiskers thanks to well-developed piloerector muscles, and during the killing bite the whiskers are usually directed forward, enveloping the bitten area in a sensitive net of hairs (Leyhausen 1979). Given the additional risks imposed by fragile sabers, improved perception would be a useful trait for the sabertooths.

Functional Anatomy: Overview

Early Interpretations of Sabertooth Predatory Behavior

One of the greatest mysteries surrounding sabertooths concerns how they used their namesake canines during the hunt. Our intuitive assumption is that the sabers must have served as formidable weapons, inflicting terrible wounds to the prey, but when we come to the question of precisely how this happened, complications arise. For several reasons, it would be impossible for the more extreme sabertooths to bite in the same way that a "normal" carnivore does: in order to get enough clearance between upper and lower canines for the bite, the jaws needed to open at huge gapes, but then the jaw muscles get so stretched that they could not gather the enormous forces necessary to drive the sabers into the flesh of prey. Various hypotheses about the paleobiology and evolution of the most popular sabertooth, *Smilodon*, were proposed during the latter part of the nineteenth century. Some scholars proposed that the animal attacked its prey hitting downward, with its mouth closed and only the tips of its canines protruding. Others thought that it used its sabers like can openers against the giant armadillo-like glyptodonts. Still others claimed that the sabertooth was the victim of a blind, irreversible "trend" in evolution toward an enlargement of the canines (Cope 1880): according to this view, in the first few million years of machairodontine evolution, the enlarging fangs might have seemed like fun (so to speak), but by the time of *Smilodon*, things had gotten out of hand, with the creature hardly being able to feed and finally going extinct—none too soon.

Around the beginning of the twentieth century, the American paleontologist W. D. Matthew conducted the first serious study of the anatomy and adaptations of *Smilodon* and several earlier sabertooths, and in 1910 he proposed the "stabbing" hypothesis. In his view, the sabertooths would not deliver a typical killing bite, in which upper and lower canines work in opposition. On the contrary, the jaw would open at an enormous gape in order to keep the lower teeth out of the way, and the huge force needed to sink the sabers into the victim would be provided by the muscles of the neck, which would move the head in an arc, using the inertia of that motion to add to the stabbing action. The whole action would be similar to the motion of the arm of a person stabbing with a knife (figure 4.17). Matthew further noticed that the mastoid process, where the cleidomastoid muscles were inserted, and the transverse processes of the neck vertebrae, where the scalene muscles were inserted, were especially developed in sabertooths, suggesting in his view an emphasis on the powerful downward flexion of the neck and head.

But other scholars objected to this hypothesis. In 1940 the Swedish paleontologist B. Bohlin claimed that the elongated, flattened shape of the sabers made them inherently fragile, so that they would most likely break under such a violent use as Matthew proposed. Taking Bohlin's arguments in mind, the famous American paleontologist G. G. Simpson, conducted new analyses, and in 1941 he refined Matthew's stabbing theory. Simpson proposed that the inertia of the predator's weight as it jumped on its prey would be added to the force of the stabbing, creating a scenario of spectacular violence. Bohlin remained unconvinced and published his objections in 1947, and many paleontologists, especially in Europe, adhered to his view. Due to the difficulties in understanding the function of the upper canines, the role of sabertooths as predators was put into question. Bohlin actually thought that they would be exclusively scavengers, unable to kill their own prey, and that they would use their canines to cut through the soft flesh of rotting carcasses. Such habits were believed to explain the abundance of *Smilodon* fossils in the tar pits of Rancho la Brea, where sabertooths would have gathered to scavenge while more efficient hunters like lions wisely kept their distance from

4.17. A human hand wielding a knife (top) to show the analogy with the stabbing hypothesis, illustrated here with the reconstructed life appearance of the head and neck of *Smilodon fatalis* (bottom). Just as the hinge of rotation for the stabbing arm is far back from the knife-wielding hand (at the elbow, and also at the shoulder), the hinge for the supposed stabbing action of the *Smilodon* head would be as far back as the cervical-thoracic joint. It was also suggested that the inertia of the predator's leap toward the prey would be added to the stabbing action.

the deadly traps (Merriam and Stock 1932). It was also suggested that the big canines would be used as a display element to assert the animal's dominance around carcasses, intimidating potential competitors.

Such hypotheses of a scavenging diet were derived from the difficulties in understanding sabertooth action in the absence of any modern analogue but were quite insufficient to explain the animals' behavior. On the one hand, it is too simplistic to assume that a fossil carnivore must have been a scavenger just because we don't know how it could have hunted. Real scavengers, such as hyenas, are not merely carnivores that cannot catch live prey; they are specialized animals with their own set of adaptations, including the abilities to process bones with their hyper-robust premolar teeth and to cover enormous distances in search of such a scattered resource as carrion is, with little energetic cost thanks to their dog-like limbs—adaptations completely absent in sabertooths. But even hyenas actively hunt a variable proportion of what they eat, and no terrestrial mammal appears to be a pure scavenger. Vultures, whose almost effortless gliding allows them to patrol enormous areas in search of carrion, are much better adapted to such a lifestyle. With their heavy limbs, retractable claws, and cutting dentition unsuited to process any food other than meat, sabertooths were ill equipped for anything beyond the occasional, opportunist scavenging that we see in modern big cats.

On the other hand, turning the canines into display objects would be extremely inefficient in evolutionary terms, neutralizing the main hunting weapons of any carnivore; furthermore, extraspecific displays would be of little use against serious competitors (such as the American lion or a pack of dire wolves, in the case of *Smilodon*) because, as modern African predators demonstrate, disputes around kill sites can become quite violent, and each competitor must be ready to turn displays into real attacks. Furthermore, the fragile sabers are quite unsuited to dirty skirmishes with other predators: they look more like delicate, precision weapons. This does not mean that sabertooths would not use their canines for display—most carnivores do. But it is a very different thing to suggest that display was the main, or even one important, factor driving the evolution of the sabers. Big mammalian predators literally make a living from their fangs, and it would simply be absurd to jeopardize their function for the benefits of display.

With these considerations in mind, it seems obvious that the sabers were primarily hunting weapons, so the big question is how they functioned. In order to solve this riddle, we first need to interpret the functional morphology of sabertooths and obtain a satisfactory picture of their biomechanics. In the sections that follow, I will deal primarily with the morphology of felid sabertooths because comparing them with the well-known, living big cats provides some of the most useful insights for understanding sabertooth action. But the other groups of sabertooths will be considered also, and discussed in detail when anatomical differences are especially significant.

The Masticatory Apparatus

The most obvious indications of how a fossil mammal dealt with its food come from its teeth. As we have seen, the teeth of derived sabertooth felids differ from those of modern cats in several ways: the incisors are large, protruding, and arranged in an arc; and the lower canines are somewhat reduced and tend to function as part of the lower incisor battery. Behind the canines, the carnassial teeth tend to become long, impressive blades with loss of internal (or lingual) cusps, and the rest of the cheek teeth may be reduced or lost.

The functional implications of these differences are complex. The enlarged, procumbent incisor battery makes it easier to hold prey, partly releasing the sabers from that function. In modern cats, the grasping action of incisors is deemphasized, while the strong, conical-shaped canines can take the impact of the first contact with prey without much trouble. In contrast, the incisors of dogs and hyenas are more similar to those of sabertooths, even though the canines of these modern animals are strong and conical like those of felines. What is the reason for such a difference? The answer is not in the mouth, but in the forelimbs of the predators. Cats use their front paws and retractable claws to hold prey and control it while they deliver the killing bite, but dogs and hyenas have narrow, inflexible forepaws that are adapted for sustained running at the cost of losing prehensile abilities—thus, their incisors play a greater role in catching and holding prey.

Sabertooth cats, as we know, had powerful forelimbs like modern cats, but they needed to protect their relatively fragile sabers from danger as far as possible, so like dogs and hyenas they emphasized incisors for use in catching prey. In documentary films we often see lions holding to their still running prey with both their claws and their canines, but that is something that sabertooths would rarely do. During initial contact with prey, they would probably restrain from biting when possible, but when they did bite fleeing or struggling prey it would be a sort of nibbling bite with their incisors and lower canines. Later, when the actual killing bite was delivered, the oversize incisors would demonstrate another advantage: once the canines sank deep enough in the flesh of prey, the incisors caught hold too and stabilized the bitten area, reducing the risk of lateral stress on the sabers. Some early sabertooth felids, like *Machairodus aphanistus*, probably paid a price for their lack of a protruding incisor battery. With their small, feline-like incisors, these creatures probably used their canines more frequently to aid in initial prey apprehension, much as lions and tigers do, but all too often their sabers broke under such stress, as shown by the high number of sabers broken in life that were discovered among the *Machairodus* sample from Batallones. Compared to the *Smilodon* sample from Rancho la Brea, broken sabers in Batallones were almost an epidemic. Although skulls of *Nimravides catacopis* (the American equivalent of *M. aphanistus*) are nowhere found in such numbers as the Batallones cats, the composite sample housed at the American

Museum of Natural History shows enough broken canines to at least suggest a similar situation. Interestingly, the sabertooth *Homotherium* of the Pliocene and Pleistocene, which developed slightly dog-like forepaws with reduced, less retractable claws (except for the huge dewclaw), also had one of the most impressive incisor batteries of all felid sabertooths, a trait that would have compensated for the partial loss of prehensile ability of the paws. A somewhat different morphology is seen in the American homotherin *Xenosmilus hodsonae*, where the incisors are very large and arranged in a marked arch, but the space between the upper third incisor and the upper canine is reduced, at least in comparison with other sabertooth cats, so that the incisors and canines almost appear to be part of a single functional unit. This arrangement led the authors who described the species (Martin et al. 2000, 2011) to coin the term "cookie-cutter cat" for this animal, and to infer for it a special type of killing bite. In their view, *Xenosmilus* would use its whole anterior dentition as a unit to puncture around a large chunk of the prey's flesh and then tear it off, causing a lethal shock. This scenario is reminiscent of a shark's bite and suggests a more indiscriminate type of attack in contrast with the precise killing bite of most other felids, which would target the throat of prey. But indiscriminate bites fit better with the robust, conical canines of hyenas and dogs (the kind of carnivores that practice that sort of biting today) than with the flattened canines of *Xenosmilus*. An alternative interpretation is that the main functional advantage of the arrangement of the incisors in *Xenosmilus* is the additional stability around the area bitten by the canines during a canine shear-bite, and that the possibility of indiscriminate bites at the body of moving prey was facultative but marginal. The morphology of *Xenosmilus* can be seen as an extreme expression of the trend in homotherins to enlarge the incisors and arrange them in an arc, but its precise functional interpretation requires further study.

The incisors of the marsupial sabertooth *Thylacosmilus* are, of course, a case apart. There is no evidence of upper incisors and only a pair of diminutive lower ones; the lower canines are small, although evidently functional – as shown by their wear facets, which indicate lateral contact with the sabers and apical (meaning on the tip or apex) contact with something else. Unfortunately, the premaxilla is broken in all known specimens, so we don't even have evidence of the upper incisors' alveoli. The space between the upper canines is so narrow that there isn't much room for incisors anyway, but it remains likely that at least a pair of small teeth existed, on account of the apical wear on the lower canines. In any case, it is obvious that *Thylacosmilus* had to use its sabers without the benefit of a strong incisor battery, indicating a different mechanism than that seen in its placental counterparts. Even the gorgonopsian pseudo-sabertooths had a strong incisor battery, although their canines had oval cross sections and were not as fragile as those of most mammalian sabertooths.

An additional way for sabertooths to protect their sabers against torsion during the bite is to enlarge the contact surface between the canine crown and the gum, or gingiva, a feature that has been inferred for

Smilodon from the extension of the cementum beyond the root-crown contact on the cervix of the upper canines (Riviere and Wheeler 2005). Such a configuration would provide several functional advantages: the gingival component of the periodontal ligament would provide additional stability to the tooth; and, since the gingiva also has a tactile function, it would help the animal to know when the tooth had reached maximum functional penetration.

The shape of the upper canines themselves also has complex functional implications. Their flattened section, which renders them so inconveniently fragile, also makes them much more efficient for penetrating the flesh of prey. The canines of *Barbourofelis* and some other sabertooths even have grooves along their crowns resembling the so-called blood grooves of soldiers' bayonets, a feature that has been wrongly interpreted as allowing the canines to be removed more easily from the wound (Diamond 1986), but whose real advantage probably has more to do with making the teeth lighter while maintaining their strength. Their curvature is best suited to penetrate along a curve that coincides with a center of rotation located outside the skull, suggesting a motion that involves not only the closing of the mandible but also some downward motion of the head. In the case of *Thylacosmilus*, the roots of the sabers remained open through adult life, indicating continuous growth – a sort of partial insurance against canine tip breakages.

The extreme blade-like shape of the carnassials and their development at the expense of the premolars indicates that meat was the overwhelmingly dominant part of the animals' diet; they were even less able to crush bones than modern big cats. It is interesting to note that, among modern cats, the cheetah displays carnassial teeth that are most similar to those of sabertooths, with reduced lingual cusps, and cheetahs are the least likely of modern cats to scavenge. Besides, they rarely have time to consume any bones of their prey, because they often have to leave their kills early to competitors. Besides cutting pieces of flesh off the carcasses, the carnassials also have the function of cutting the skin of prey, both for consumption and to gain access to the flesh underneath. The incisors may add in the first function, but modern carnivores rarely use them to cut skin, and sabertooths probably behaved the same way. So the relative length of the carnassials is likely to be a measure not so much of how much meat a cat can eat, but of how efficiently it can open and process a carcass. From this point of view, the huge carnassial blades of *Homotherium* or *Barbourofelis* suggest fast, competitive, almost frenzied feeding, probably associated with group living, relatively open habitats containing competing carnivores, or both. *Megantereon*, on the other hand, had relatively small carnassials for a sabertooth, and we can imagine it leisurely consuming its prey in the branches of a tree or in the shade of deep bush, aloof and solitary like modern leopards or jaguars. *Thylacosmilus* had no true carnassial teeth, but all the cheek teeth were elongated with the inner cusps reduced, forming what was almost a continuous, vertical shearing blade all along the post-canine row. This is the most extreme adaptation to hypercarnivory among South American marsupials.

The anterior joint between the two halves of the mandible, or symphysis, is also different in sabertooths and non-sabertoothed carnivores. In the former, it is vertically reinforced and often enlarged to form a mental process of varying length, sometimes encompassing the whole length of the sabers. This reinforcement argues against the notion of a weak bite and of the lack of functionality of the lower teeth during the killing bite. On the contrary, the strong symphysis suggests that the anterior part of the mandible withstood strong vertical forces during the bite, although it would have been vulnerable to random lateral stresses – a vulnerability that required prey to be kept as immobile as possible. Related to this is the nearly vertical orientation of the sabertooth's lower canines when seen from the front, in contrast with their V-shape orientation in modern cats. The former is again best suited to resist vertical forces, while the latter is better able to stand lateral stress from struggling prey.

The joint between the mandible and the skull shows several adaptations for large gapes. The articulation itself is in a low position relative to the palate and the occlusal plane of the teeth, so that any rotation around it will produce a greater aperture. The concavity of the glenoid process is relatively shallow and allows for the rotation of the mandible along a wider arc than in modern cats.

The insertion areas of the main adductors of the mandible (or jaw-closing muscles) in sabertooths demonstrate that they were adapted to wide gapes, too. The low coronoid process implies the presence of a longer temporalis muscle, and the high sagittal crest contributes to the same effect. Muscle fibers contract optimally when the extended length does not exceed one and a half times the contracted length, and in order to keep that proportion during gapes of up to 100 degrees, it is necessary to increase the contracted length of the fibers. In spite of such adaptations, the bite force of the temporalis and masseter muscles was relatively smaller in sabertooths than in a normal cat at the largest gapes, but it increased rapidly as gapes decreased, so that the force exerted during the carnassial bite (when the gape is already small) was quite strong (Bryant 1996a). The relatively posterior position of the carnassials, which were thus closer to the axis of rotation of the mandibular joint, also helped increase the force of the bite.

The Neck and Head Muscles and the Canine Shear-Bite

The differences between mammalian sabertooths and non-sabertoothed carnivores in the areas for muscle insertion in the back of the skulls and in the cervical vertebrae have very important implications for understanding the mechanism of the sabertooths' killing bite. The first striking features we notice are the changes in the mastoid region: the anterior and ventral projection of the mastoid process in sabertooths is usually associated with a retraction of the paroccipital process, a condition opposite to that seen in most modern carnivores, creodonts, and non-sabertoothed marsupial carnivores. The changes in the paroccipital process are related to the need to open the jaws at wide gapes, because the main jaw-opening muscle, the digastric, is inserted into that process. Retracting the process tip allows for more space between the origin and the insertion of the muscle, which then can rotate the joint around a wider arc while maintaining the right proportion between contracted and extended length.

But the changes in the mastoid process have different implications. First Matthew (1910) and then Simpson (1941) quickly noticed that the mastoid included insertions of muscles that pull the skull down, and they particularly identified a group of muscles including the brachiocephalic and the cleidomastoids, which pull the head against the sternum and upper arm. They incorporated this evidence into their stabbing hypothesis, because if the insertions for these muscles were enlarged, it appeared obvious that pulling the head down from the shoulders was essential for sabertooth action. But that observation explained only part of the truth. The scarcity at the time of good descriptions of the musculature in modern carnivores helped hide important elements from those early paleontologists.

In 1985 the American paleontologist W. Akersten used the detailed description of muscle insertion areas in the skull of a giant panda done by D. Davis in 1964 as a guide to interpret the morphology of the mastoid area in *Smilodon*. Akersten found that the brachiocepahlic and cleido-mastoids were actually inserted into a thin band on the external part of the mastoid process, but its main area was occupied by the extensive attachments of muscles that originated in the underside of the wings of the atlas, or first cervical vertebra. Cautioning that the homology of elements between the sabertooth and the distantly related panda could be incomplete, Akersten transferred the pattern to the fossil, which had striking implications for the interpretation of the biting mechanism.

If the atlanto-mastoid muscles (those extending between the atlas and mastoids) were the most directly affected by changes in mastoid morphology, then the depression of the skull around the articulation with the atlas, and not relative to the shoulder or anterior trunk, was the key action that marked the difference between normal carnivores and sabertooths. Years later, our own dissections of big cats (Antón et al. 2004c) showed that there were some differences in detail between the muscle insertions of the giant panda as described by Davis and the pattern present in the living relatives of *Smilodon*. However, the essential fact remains: it is a muscle running from the atlas to the mastoid–specifically, the obliquus capitis cranialis, which occupies the main area of the mastoid. Akersten also noticed that the posterior elongation of the wings of the atlas, so typical of sabertooths, quite precisely matched the transformation of the mastoid. Evidently, the atlanto-mastoid muscles in sabertooths had longer fibers and were able to rotate the atlas-skull articulation along a wider arc than was possible for other carnivores.

4.18. A sequence of drawings showing the scimitar-tooth *Homotherium* applying the canine shear-bite to a prey. Top: the jaws are wide open, and the lower canines and incisors get anchored against the prey's body; center: the whole head is pulled downward by the anterior neck muscles, driving the sabers' tips into the flesh; bottom: once the jaw closure advances and the gape is small enough, the muscles of mastication bring the jaws closer together. A backward pull of the predator's head would further enlarge the wound.

From these observations, Akersten built a hypothetical model for the killing bite of *Smilodon* called the canine shear-bite, which differed substantially from the stabbing hypothesis. In this scenario, the killing bite began as the predator opened its jaws at full gape against a conveniently convex section of its victim's body, such as the throat or, as favored by Akersten, the skin fold between the thigh and the belly. Since the masseter and temporalis muscles were in an unfavorable position to exert enough force for the penetration of the upper canines, at this point the head depressors, particularly the atlanto-mastoid muscles, went into action, pulling down the skull and aiding the penetration of the saber tips. Meanwhile, the lower jaw worked as an anchor, with a fixed position in the flesh of prey, against which the head depressors could act. Once the jaws had closed to a sufficiently low angle, the masseter and temporalis could exert full pressure and put additional force into the bite. One final motion hypothesized by Akersten was a backward jerk that would result in the tearing of a whole chunk of flesh from the prey's body (figures 4.18 and 4.19).

The canine shear-bite hypothesis elegantly solved many of the problems of the stabbing scenario, but it left some questions open. The main muscles involved in the stabbing hypothesis (the brachiocephalicus and associated muscles and the scalene group) were given secondary roles in the canine shear-bite, contributing additional force to the downward motion of the head. But since the killing bite was now supposed to occur from a static start, with the head attached in a fixed position by the anchoring of the mandible on the prey's body, the contribution of muscles that tend to rotate the whole neck and head around a far posterior axis were difficult to integrate into the scheme. If the main business of the killing bite is restricted to the area ahead of the axis vertebra, why not retain a shorter neck with more conventional muscle insertions, instead of evolving a long and muscular one that costs more energy to maintain?

The exceptional sample of skulls and cervical vertebrae of *Homotherium* from the Spanish site of Incarcal shed some light on these problems. While studying the Incarcal fossils some years ago, the Spanish paleontologist A. Galobart and I noticed that the cervicals differed from those of modern cats in more ways than Matthew and Simpson had remarked for *Smilodon* (Antón and Galobart 1999). As both Matthew and Simpson had pointed out, the processes into which the scalene muscles were inserted indeed projected more ventrally than in modern cats, but we observed that the whole transverse processes, including insertions for a range of other neck muscles, were more developed and projected laterally. Some processes actually showed a marked dorsal extension. In addition, each individual vertebra of *Homotherium* was relatively longer than the equivalent element in a modern cat, a feature that had already been observed in several sabertooth species. These observations have important functional implications. The greater length of the neck allowed for a wider range of rotation of the head, and the enlarged transverse processes allowed several muscles to produce wide lateral, ventral, and dorsal rotation of the neck, and to maintain any position with considerable strength. We

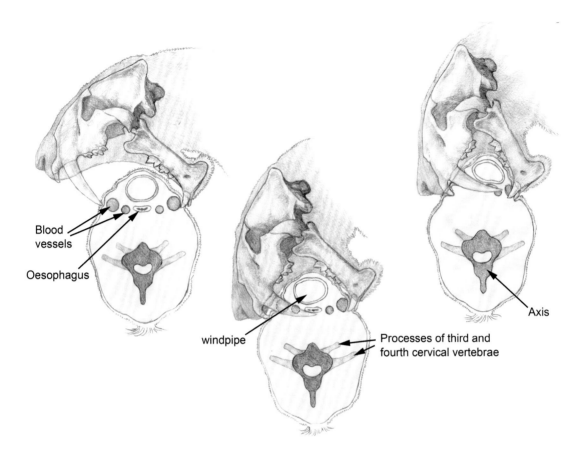

Blood vessels

Oesophagus

windpipe

Processes of third and fourth cervical vertebrae

Axis

did not find the exclusive emphasis on ventral flexion of the neck and head that would be expected in a stabbing scenario, and it appeared that the posterior neck had a greater role in the bite than initially proposed in the canine shear-bite hypothesis. In the resulting scenario, *Homotherium* would use its long neck to position the head in a very precise orientation for the killing bite, and then the various muscles would be able to hold that precise position quite strongly against any struggling motions of the prey. This combination of strength and precision fits well with the idea of a canine shear-bite, but the need for precision seems more adequate for a throat bite than for a belly bite. After all, a longer neck is a structurally weaker neck unless lots of additional muscle mass are added, which is an energetically costly adaptation, so the additional reach and precision of the sabertooth neck must have been an important advantage (figure 4.20). Modern hyenas, for instance, have evolved long necks for very different reasons than sabertooths, and they certainly don't have the extra development of lateral and ventral musculature or the associated projecting transverse processes. On the contrary, they concentrate their muscle mass on the dorsal region of the neck, where the extensor muscles, necessary for the carrying of large carcass pieces, are located (Antón and Galobart 1999). A study of the mastoid anatomy in the early homotherin *Machairodus aphanistus* from Batallones (figure 4.21) shows that it had

4.19. A sequence of drawings showing the sabertooth *Megantereon* biting at the neck of a horse, from beginning (left) to end (right). The section through the horse's neck shows the position of the axis vertebra, the main blood vessels and the trachea. The narrowing of the neck at the throat, ventral to the anterior cervical vertebrae, offers an ideal anchoring point for the bite, and as the sabers penetrate the flesh of the prey, they are very likely to damage one or more of the major blood vessels. The larger the saber, the greater the possibility of cutting the vessels and causing a deadly wound.

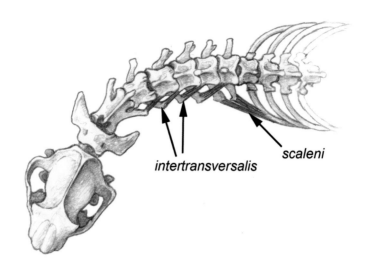

4.20. *Homotherium latidens* seen from above as it bites the neck of a horse. Top: anatomical details of the neck. Note the trajectories of the intertransversalis and scalene muscles, which provide a stronger and more precise rotation thanks to the great lateral projection of the cervical transverse process, where they attach. Bottom: reconstructed life appearance of the predator and prey.

intertransversalis

scaleni

a morphology more similar to that of a pantherin cat than to that of *Homotherium*, but even so, it already showed incipient adaptations for the canine shear-bite, including the development of posteriorly projected atlas wings (Antón et al. 2004a, 2004b).

The development of a long, strongly muscled neck is taken to the extreme in the marsupial *Thylacosmilus*, whose cervical vertebrae have not only impressive transverse processes, but also prominent ventral keels associated with the presence of powerful longus colli muscles, which are ventral flexors of the neck (figure 4.22). In correspondence with this feature, *Thylacosmilus* also displays marked insertions in the ventral side of the skull for the longus capitis, a powerful flexor of the head that is a continuation of the longus colli (Argot 2004). We found similar, if less

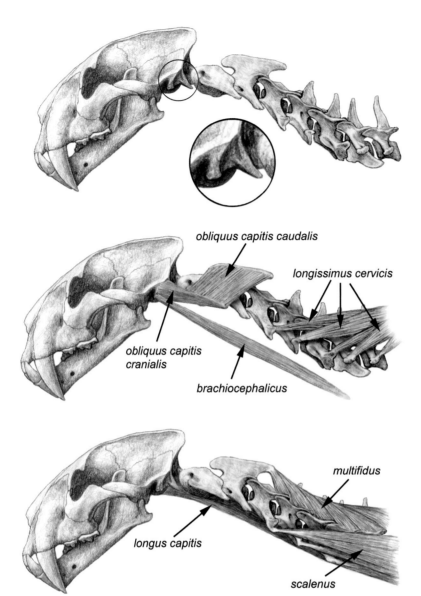

4.21. Skull and neck anatomy of *Machairodus aphanistus.* Top: skull and cervical vertebrae, with detail of the mastoid region enlarged in the circle. The color code corresponds to the attachment areas of different muscles, with that of the brachicephalicus in green, obliquus capitis cranealis in blue, and digastricus in orange. Center: deep muscles, including the lateral depressors of the head attaching to the mastoid process. Bottom: deep muscles, including a deep depressor of the head, the longus capitis, and others involved in lateral and vertical movements of the neck.

obliquus capitis caudalis

longissimus cervicis

obliquus capitis cranialis

brachiocephalicus

multifidus

longus capitis

scalenus

pronounced, marks in the Incarcal skulls of *Homotherium*, indicating that these ventral muscles also contributed to the motions of the head involved in the canine shear-bite in this placental sabertooth.

An additional feature of the sabertooth skull related to the involvement of neck muscles in the biting action is the elevation of the occiput observed in many unrelated species, including the felid *Smilodon populator*, the barbourofelid *Barbourofelis fricki*, and the marsupial *Thylacosmilus atrox*. In more primitive, typical carnivores, the occipital plane is more inclined, even close to the horizontal, and the unusual condition observed in advanced sabertooths suggests that these animals had a greater ability to elevate the whole head relative to the neck. Thanks to that morphology, the muscles pulling the top of the skull backward from the neck

4.22. Sequential reconstruction of the neck and head of *Thylacosmilus*. Top: skull mandible and cervical vertebrae; middle: musculature; bottom: reconstructed life appearance.

could pull along a wider arch as the skull rotated around its articulation with the atlas (Martin 1980). Elevating the whole skull makes sense in terms of the large gapes of sabertooths, both to help open the mouth in anticipation for the bite and to provide the whole head with a greater arc of rotation for its downward thrust during the canine shear-bite.

Body Proportions, Musculature, and Locomotion

The forelimbs of carnivores are subject to conflicting functional demands, since they need to perform both locomotory and predatory actions. Efficient terrestrial locomotion is best achieved with light limbs, with elongated distal segments (to increase the length of each stride), muscles concentrated on the proximal part of the limb (to reduce the weight of swinging segments and thus the effort necessary to move the limb), and articulations that restrict rotation to the vertical, or sagittal, plane. Holding prey, on the other hand, requires almost the opposite design: broad limbs with short and well-muscled distal segments that provide great strength, and flexible articulations that allow lateral rotation, pronation, and supination of the manus and forearm – forelimb features that are largely similar to those involved in climbing.

Starting from an ancestral condition that combined small size with at least some climbing adaptations, the different carnivore groups have evolved different solutions to this dilemma. Dogs and hyenas have developed long, cursorial limbs, and the function of holding prey is largely transferred to the anterior dentition. Cats in general have retained some of the climbing adaptations of their ancestors, including flexible wrists and forearms, well-developed forearm musculature, and retractable claws, and have put them to use in catching prey. One partial exception is the cheetah, which has lost much of the strength and flexibility of its forepaws, as well as part of the retractability of its claws, in order to gain speed. The cheetah partly compensates by having an unusually large dewclaw, which it uses to hook prey during full-speed pursuits.

As can be expected, the forelimbs of sabertooths largely conform to the cat model, actually taking it to extremes in many cases. *Smilodon*, *Barbourofelis*, and *Hoplophoneus* all had short, extraordinarily well-muscled forelimbs capable of ample lateral rotation and armed with large, retractable claws. It is difficult to realize how large and heavy the paws of a lion or tiger are until you actually hold them in your hands. While participating in dissections of big cats, I have had the opportunity to seize the paws of the animals, and I realize how a single blow from those paws can demolish an adult human almost effortlessly. Considering that the paws of *Smilodon populator* were considerably broader and heavier than those of any living cat, and they were powered by a much more muscular arm, it is frightening to imagine the devastating effects of a blow from such paws. It is obvious that the more robust sabertooths took no chances with the struggles of their prey and their potential damaging effects on their fragile sabers. A medium-sized ungulate caught under the weight of *Smilodon* was rendered as effectively motionless as if buried under tons of rocks.

The lack of retractable claws in the marsupial *Thylacosmilus* has usually been seen as a handicap for such a specialized sabertooth, whose sabers were as elongated as in the most extreme of placental sabertooths. How did it manage to subdue its prey without the aid of catlike claws? In fact, *Thylacosmilus* is an excellent example of the way evolution works from the available raw materials. Partially retractable claws are probably an ancestral condition of all placental carnivorans, so the development of full retractability is a relatively simple evolutionary step for an animal

4.23. Reconstructed life appearance of *Promeganterreon* climbing a tree. The body proportions of this early smilodontin sabertooth–with a long and flexible back, long hind limbs, and forelimbs with good rotation and grasping capabilities–suggest it was a very able climber.

with the right ancestry (and not too far removed from the ancestral condition). The ancestral forms of the South American carnivorous marsupials, however, show not the slightest trace of claw retractability, and it would be quite surprising (although not impossible) for *Thylacosmilus* to evolve that feature. The remarkable similarities between *Thylacosmilus* and placental sabertooths show the evolutionary plasticity of the mammal body, but the differences in detail are just as important for demonstrating that there are limits to what adaptation can do against genetic constrain. Nonetheless, the structure of the forelimbs of *Thylacosmilus* indicates enormous muscular strength, with especially powerful deltoids and pectorals; and broad paws capable of considerable lateral rotation, including a partly opposable thumb–so the animal was well able to use them for subduing prey. Modern bears also lack retractable claws, but you don't want to be caught between the paws of even a small species of bear–it can be a veritable embrace of death.

Some sabertooths, however, have developed cursorial traits in their forelimbs. Members of the genus *Homotherium* from the Pliocene and Pleistocene are the most extreme example, but earlier members of the lineage already show signs of that adaptation. The forearm of *Homotherium* was relatively long, with a narrow distal articular area that in turn corresponded with a narrow wrist, less capable of lateral rotation than that of a smilodontin or even a tiger. The claws of *Homotherium* were relatively small and less fully retractable than those of either modern cats or smilodontins. A similar pattern is already apparent in the late Miocene *Lokotunjailurus*, and to some degree in *Amphimachairodus* and *Machairodus*, all of which had a relatively huge dewclaw–a feature all the more striking because of the moderate size of the other digits' claws (see figure 3.45). This pattern clearly resembles that found in the modern cheetah, suggesting that the homotherins sacrificed part of their forepaws' grasping power in favor of more efficient terrestrial locomotion. Even the other, less cursorially adapted, machairodonts also had disproportionately large dewclaws.

The hind limbs of felid sabertooths show their own patterns of specialization. In the Miocene genera *Promeganterreon*, *Machairodus*, and *Amphimachairodus*, they resemble the hind limbs of pantherin cats in general morphology and proportions, although in general the claws tend to be small. The size is probably a response to increased terrestrial locomotion: large, retractable claws are useful for frequent climbing, but smaller, less retractable ones make for a lighter foot and can contribute a little more to traction during locomotion on the ground, like the spikes in an athlete's shoes. In the Pliocene, *Megantereon* developed a relatively more robust hind limb than *Promeganterreon* (Adolfssen and Christiansen 2007) but retained a similar morphology, indicating an overall similar "scansorial" locomotion. *Homotherium*, while becoming more long-legged and cursorial in its overall anatomy, also developed a series of bear-like morphological features, which as discussed above were interpreted by some as indications of plantigrade posture, but which more

likely point to the increased stability of the hind limb while subduing struggling prey.

The lumbar region of the vertebral column also changed in derived sabertooth felids. Miocene genera had long lumbar vertebrae, corresponding with long, flexible backs similar to those of modern felids. Such backs fit well with the scansorial habits of most felids, as flexion and extension of the back contribute to the motions of the cat when climbing and permit sudden acceleration when the stalking animal accelerates suddenly from cover (figure 4.23). Pliocene and Pleistocene felid sabertooths display shorter lumbar vertebrae, more like those of dogs, hyenas, or even ungulates, which indicates rigid backs that transmit the propulsive energy of the hind limbs in a passive manner. If *Homotherium* needed to save energy when covering long distances on land, a short and rigid back would be advantageous. But such a short back is also stronger, and it would be a useful adaptation when struggling with prey. Thus, it is also found in sabertooths that would not travel as extensively, such as *Smilodon* and *Barbourofelis*. The latter genus has what could be the most derived lumbar vertebrae of any placental sabertooth, resembling the vertebrae of a badger more than those of a cat. The badger needs the support of a strong, rigid back to keep its hind feet firmly planted on the ground while it uses its enormously powerful forelimbs for digging; similarly, the barbourofelids (like all sabertooths) would have benefited from a strong back to keep their place on the ground while holding their struggling prey.

However, it should be noticed that, although the lumbar vertebrae of many sabertooth species are articulated in a way that greatly limits lateral motions, vertical flexion remains possible in all groups, so that the back is still able to contribute substantially to the gallop via flexion and extension in the sagittal plane. This ability is important for quick acceleration from cover, and in the modern cheetah the flexion of the lumbars contributes significantly to speed (Hudson et al. 2011). It has been said that such motions would allow the animal to move at some ten kilometers per hour, even without its limbs (Hildebrand 1959, 1961)!

In summary, many of the morphological features of derived felid sabertooths point to a combination of an increasingly efficient terrestrial locomotion compared to that of feline cats, and greater strength, necessary in order to subdue struggling prey. Different groups of sabertooths have emphasized one or another of these adaptations: for instance, homotherins developed a more fleet, cursorially adapted body plan, and the smilodontins developed a more hyper-robust, wrestling physique. The smilodontins were not alone in developing such features, and their similarities with the robust homotherin *Xenosmilus hodsonae* or the powerful barbourofelid *Barbourofelis fricki* are striking. All these animals developed short, well-muscled legs, and it is obvious that they favored strength over speed when dealing with prey (Wroe et al. 2008).

In the case of the marsupial sabertooth *Thylacosmilus*, several anatomical features of its hind limbs and back show that, like its placental

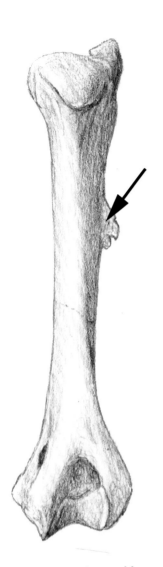 

4.24. Pathological humeri of sabertooth felids in posterior view. Left: *Homotherium latidens,* from Senèze, France; right: *Smilodon fatalis* from Rancho la Brea, United States. Both specimens show pathological bony growths in the area where the acromial part of the deltoid muscle was attached to the humerus. The growths correspond to attempts of the organism to repair damage after repeated strains caused tears in the muscle. Figure 4.27 shows the action of the deltoid muscle that could lead to such an injury during a hunt.

counterparts, it was able to use great force in wrestling down its prey. Its lower back was more rigid than that of its non-sabertoothed relatives, the articular head of the femur reveals a greater postural flexibility, and the hind feet were semiplantigrade—features that we are familiar with after our review of placental sabertooth osteology and that indicate the ability to pull the prey down while standing with feet firmly planted on the ground (Argot 2004).

Paleopathology

Injuries and trauma in fossil bones offer rare insights into the hazards of an extinct animal's life. The best sample of pathologic sabertooth bones comes, of course, from Rancho la Brea, where there is a striking abundance of bones showing evidence of traumatic injuries. One common type of pathology is the presence of bony growths on the area corresponding to the insertion of the deltoid muscle, in the proximal

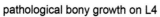
pathological bony growth on L4

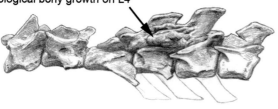
neural process of L3

mamillary process of L5

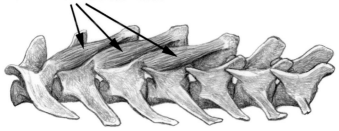
bundles of the *multifidus* muscle

half of the humerus (figure 4.24). This feature indicates that the deltoid was repeatedly strained to the point of tearing during forceful attempts by the animal to flex its shoulder, such as would occur while the saber-tooth was pulling a heavy prey animal toward itself. A similar pathology was found in the complete skeleton of *Homotherium* from Senèze, in France (Ballesio 1963) and, more recently, in a complete humerus of the same species found in Saint-Vallier, also in France (Argant 2004). Such frequency supports the idea that the origin of this trauma lies in a habitual activity, such as hunting. Traumatic injuries in the sternum are also frequent in *Smilodon* from La Brea, suggesting that the animals often collided violently, chest first, with large prey. Bony growths on the lumbar vertebrae also testify to the frequent tearing of the spinal muscles during struggles with heavy prey. One example of a similar pathology is also found in a partial skeleton of *Megantereon* from Kromdraai, in South Africa (figure 4.25).

Fractures of limb bones have also been found, testifying to accidents, and crushed feet also suggest that large prey could occasionally trample a cat's comparatively fragile foot.

Other examples point to aggression between predators. One skull of *Smilodon* from Rancho la Brea displays a wound on its forehead perfectly matching the shape and size of a *Smilodon* saber, and in view of the lack of substantial healing of the bone around the hole, it seems that the injury was fatal (Miller 1980). Another discovery at the same site was a *Smilodon* scapula apparently pierced by the saber of a fellow cat (Shaw 1989).

One of the most classical examples is a skull of the nimravid *Nimravus* from the White River Oligocene in South Dakota that displayed a deep wound, most likely caused by the saber of *Eusmilus* (Scott and Jepsen 1936). The saber penetrated the frontal bone, piercing the sinus, but it did not reach the brain, and the victim survived the attack long enough for the bone to heal completely (figure 4.26).

Reconstructing the Sabertooths' Hunting Sequence

Putting together all the evidence reviewed in the previous sections, we should now be able to visualize the likely chain of actions in a sabertooth hunt. The main example will be *Smilodon*, the quintessential sabertooth, but we will also take a look at other kinds of mammalian sabertooths and the inferred differences in their hunting styles (figure 4.27). The therapsid sabertooths being so different from the mammals, I will discuss both their functional anatomy and their hypothetical predatory behavior in a separate section below.

Let us now imagine a lone, adult female *Smilodon fatalis* in the Rancho la Brea area. She has come across a nice herd of bison, and waits hidden among some bushes near the edge of a pool, hoping the herd will come near enough. Let us remember that *Smilodon* is a heavy cat and not well equipped for long-distance running, with its short distal limbs and especially the metapodials, so it must come really close to its prey to

4.25. Lumbar pathology in a sabertooth. Top: lumbar series (second to fifth lumbar vertebrae) of *Megantereon whitei* from Kromdraai, South Africa. All the vertebrae are more or less broken; the transverse processes of the third to fifth lumbars were restored on the basis of those from the left side. An abnormal bony growth (only on the right side) combines three of the vertebrae into a rigid mass and concentrates on the fourth lumbar. Center, right: a healthy lumbar series of a lion, *Panthera leo*. Center, left: the position of three bundles of the multifidus muscle in the lion's lumbar vertebrae. Note that each bundle spans two vertebrae, from the mamillary process of the more caudal vertebra to the neural process of the more cranial one. Such a disposition mirrors the growths in the fossil vertebra. Bottom: reconstructed life appearance sketch showing a situation in which the multifidus of the right side could strain and cause lesions like those in the fossil. The felid is trying to hold onto a struggling antelope, and as a consequence its back is twisted violently.

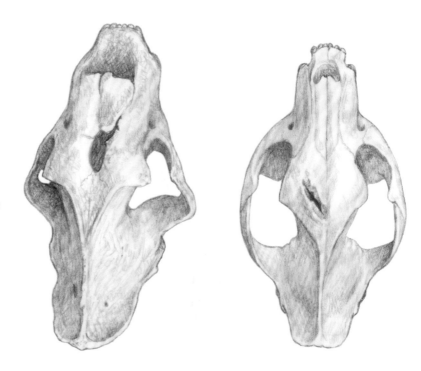

4.26. Skulls of *Smilodon* (left) and *Nimravus* (right) with saber wounds. The wound on the *Smilodon* skull was probably inflicted by a conspecific. It shows no signs of healing and was apparently fatal. The wound on the *Nimravus* skull is attributed to the nimravid *Eusmilus,* and it pierced the frontal bone and entered the sinus, but it did not penetrate the brain. It shows extensive healing, indicating that the animal survived for a long time after the conflict.

be able to catch it. After a long wait, the bison pass a few dozen meters in front of our cat, but she doesn't break from ambush. She is waiting for an appropriate prey. Finally, she detects a young bison, perhaps a yearling, and starts creeping toward her target. Inch by inch, the sabertooth approaches the unsuspecting victim until she is very close, perhaps twenty meters away.

Then the cat explodes from hiding. She covers the distance to her prey in a few long, lightning-swift bounds. Her heavily muscled limbs are not very good for sustained running, but they are capable of quite sudden acceleration from a static start, and her heels permit a good bounding gallop. The loins, although shortened, are capable of powerful and efficient ventral flexion, which contributes to the length of each bound. Actually, the cat is not running at full speed – bison are relatively slow creatures, and *Smilodon* needs its highest gears for lighter prey such as horses. While the rest of the herd runs away in panic, the cat smashes into the young bison in mid-leap, 250 kilograms of predator against perhaps 400 kilograms of prey. The young bison loses balance for a moment but manages not to fall. The impact is especially heavy on the cat's sternum, but this time there is no lesion to the bone. The predator puts her heavy paws around the prey's shoulders, plants her own hind feet firmly on the ground, and pulls. Now the sabertooth's short, heavy hind limbs and back are put to the test, and the muscles of the cat are tensed, sharply sculpted under the skin. The bison has also planted its four feet solidly, but after a violent pull of the cat, it collapses on one side. This is something that a lion, even a male, could not have done single-handedly. But even our *Smilodon* has faltered for a moment, as the stress of the pull has awakened the pain of an old injury. The pull of the deltoid during this crucial part of the hunt has strained the shoulder muscle fibers before, and the animal has a bony growth on her humerus from her body's efforts to heal the injury. This time, the pain is not too bad, and the cat keeps control of her quarry.

Now the cat quickly positions herself in a crouch beside the back of the fallen bison, puts the weight of her heavy forequarters over the prey's chest and shoulder, and pins the bison's neck down against the ground with the iron grip of one forepaw. The other forepaw maneuvers to hold the head of the animal. The extraordinary strength and capability for lateral rotation of the forepaws are essential at this point. Now the throat of the bison is reachable, but at a slightly odd angle. The cat cannot loosen its grip on the prey to adjust its position without risking the loss of the prey, but fortunately the neck of *Smilodon* is long and very strong. The sabertooth reaches with her head and turns it until the bison's throat is conveniently in front of her jaws. Then she opens her mandible at full gape, until the ventral part of the prey's throat is encompassed between the tips of her fangs. The muscles of her neck are fully tensed now, as the cat's head is pulled down violently, sinking the upper canines into the flesh of the prey, which–feeling the piercing canines–thrashes with all its strength. But the area where the cat is biting hardly moves at all, so strongly is she holding her prey.

All this action has taken place in just a few seconds, but now some members of the bison herd are returning to the intended kill site. One of the adults charges the cat, and she must let go of her prey for a moment But then she crouches defensively and snarls at the adult bison, gaining the advantage and grabbing the yearling again before it manages to rise to its feet. Blood is flowing abundantly from the victim's throat, due to the severing of a major artery. The adult bison stand hesitantly in front of cat and prey, but all the commotion, the bovine bellowing, and the rising dust have not passed unnoticed by other predators. A couple of large *Smilodon* appear from the other side and approach the scene with a confident trot. These are two young males, probably in their second year but already larger than the female, and she would not stand a chance of defending her kill against them. But she doesn't retreat, and the two newcomers hold the victim's hindquarters, one of them biting at the abdomen. In the meantime, the adult bison have turned their backs and started to flee in the presence of the new predators. The newcomers happen to be the cubs of the hunter from a couple of years before. She has bitten the throat of the young bison again, and the prey is now lying in a pool of blood, its struggles growing ever fainter. Even before it dies, one of the cubs applies a carnassial bite to the skin between the belly and the hind leg, cutting the hide and causing evisceration. Now the female is panting for breath, but the youngsters are already feeding hurriedly. As soon as her breath slows down a little, the female starts feeding quickly as well, and in a few minutes the three cats have gulped down an amazing amount of meat. None too soon: in the distance, the figures of a group

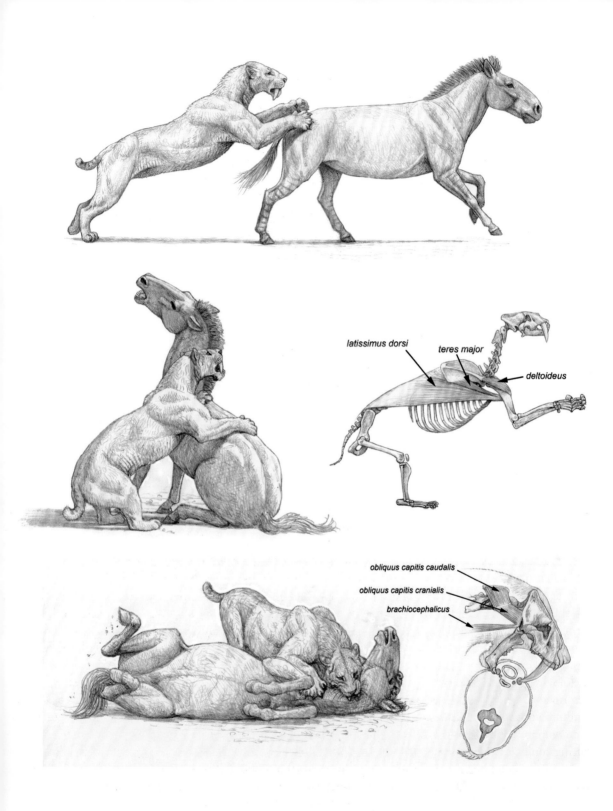

latissimus dorsi

teres major

deltoideus

obliquus capitis caudalis

obliquus capitis cranialis

brachiocephalicus

of American lions can be seen taking shape amid the heat waves. There are perhaps half a dozen lions, each animal larger than *Smilodon*. With their bellies reasonably full, the sabertooths prefer to leave the carcass to the lions rather than risk breaking their precious sabers during the fight that would ensue. It is time for a catnap in the nearby woodland.

This idealized sequence of events portrays many aspects of the typical sabertooth predatory action, especially when taking down large bovids, although in the case of the lighter homotherins it is likely that more than one individual participated in the hunt in order to add firepower and compensate for the smaller muscular power of each cat (figure 4.28). Modern lions hunt young proboscideans more often than was traditionally thought, and the fossil evidence clearly indicates that at least *Homotherium* did so regularly, too. Taking such prey was made especially difficult by the tight social system of proboscideans, and it would have required a large number of cats working together, a high sense of strategy in order to distract the attention of the adult proboscideans, or both. B. Kurtén indulged in a delightful literary recreation of such behavior in his novel *Dance of the Tiger* (1980). At any rate, the actual killing of the prey would proceed along the same lines as described above for the buffalo hunt (figures 4.29 and 4.30). Sabertooths shared their environments with a host of other large mammals such as hippopotamuses, rhinos, and giant ground sloths. In all these cases, even with the impressive weaponry of machairodontines, it is likely that the younger individuals would be favored as prey, and the adults of such giant species would remain immune to predation most of the time (figure 4.31).

Conflict at the end of the hunt would be a likely outcome whenever a conspecific was in sight, and it would not usually end so amicably as in the case of the mother and grown cubs described in our fictional narrative. Even if some species of sabertooth had some sort of group life, any individual not belonging to that limited group would be considered a rival, and fighting would ensue – with consequences that were often very violent (figure 4.32). Proof of such conflicts is found occasionally in the form of sabertooth fossils with injuries that were most likely caused by another sabertooth, as mentioned above in the section on paleopathology. Such examples of conflicts between predators also show that the use of sabers was not always of the careful, almost surgically precise kind described when discussing the canine shear-bite method. The fossil record provides evidence both of instances when sabertooths did bite on bone with their upper canines and of broken sabers. Canines could break during conflicts between predators and also during the hunt, as we have seen in the section on the masticatory apparatus in the discussion of the Batallones-1 sample of *Machairodus aphanistus*. Also, some prey would leave the sabertooths little option but to risk biting on bone. One example of this is a skull of the giant armadillo-like glyptodon, *Glyptotherium texanum*, found in Pleistocene deposits in Arizona. This skull displays a pair of elliptical holes in the frontal bones that are best explained as wounds

4.27. A sequence of drawings showing *Smilodon* hunting a horse. Top: to catch the prey, *Smilodon* rushes from cover in a short series of bounds. Its back had limited lateral flexibility but could flex and extend well enough in the sagittal plane, contributing to the length of each bound and helping the animal accelerate rapidly. Center: to wrestle down large prey, *Smilodon* takes advantage of its powerful forelimbs and its short, strong back. The acromial part of the deltoid was one of the main muscles pulling the arms back, and its attachment area in the humerus often shows signs of strain (see figure 4.24). The latissimus dorsi is a powerful muscle pulling the arms from the lumbar fascia, and if the back is short, its pull is more effective. It shares its attachment in the humerus with the teres major, which pulls back from the huge scapula, contributing additional strength. Bottom: to proceed with the kill, the cat holds the prey's head and neck in its powerful paws and delivers a precise canine shear-bite. With jaws agape and the lower incisors and canines anchored against the prey's neck, *Smilodon* pulls down its whole head with the strength of the obliquus capitis muscles, aided by the pull of the brachiocephalicus. That motion brings about the penetration of the sabers, which tear through the prey's blood vessels and cause massive blood loss.

4.28. A scene in the early Pleistocene of Spain, with a pair of *Homotherium* in pursuit of a primitive bison.

caused by the sabers of the contemporary sabertooth cat *Smilodon* (Gillette and Ray 1981). Given the heavy, full-body armor of glyptodons, a bite to the head would be about the only option for any predator intent on attacking them.

The Functional Anatomy of Therapsid Sabertooths

The anatomy of gorgonopsians is profoundly different from that of the mammalian sabertooths we have been discussing, but there are many features in their skeletons indicating that these were not sluggish reptiles, but quite active predators.

Their brains, of course, were relatively much smaller and simpler than that of any mammal. Their eyes were relatively small and laterally placed, providing only limited stereoscopic vision. It is interesting to see that gorgons had well-developed turbinals, a labyrinth-like bony structure within the nasal cavity that is associated with an advanced sense of smell. Turbinates are present in mammals but lacking in reptiles. It is obvious that the superior sense of smell would help gorgons to track their prey and also to find carrion if necessary.

The incisors of therapsid sabertooths formed an impressive arc in front of the sabers, and they clearly played a decisive role in holding prey. In these predators, there was probably less differentiation between the phases of the hunt—such as the initial contact, immobilization of the prey, and making the killing bite—than in mammalian sabertooths.

4.29. A scene in the late Pleistocene of western North America, with a group of *Smilodon fatalis* hunting a young Columbian mammoth. Although the young mammoths were relatively easy for the powerful cats to kill, the presence of the mother and other adult proboscideans made this a very dangerous hunt.

It is more likely that therapsids simply slashed at the prey, causing as much damage and blood loss as possible from the start. Feeding would be simple, because in the absence of any substantial post-canine teeth, all the cutting and tearing of flesh would have to be done by the incisors, taking large chunks off the carcass and gulping them down quite unceremoniously.

The masticatory musculature was still of a reptilian type, but with a particular set of specializations (figure 4.33). Unlike the case in mammals, the mandible of gorgons was made up of several bones, and it had a double articulation with the skull. To allow these animals to bite at large gapes, there was a special adaptation: the quadrate (one of the skull bones involved in the articulation with the mandible) was able to move in several directions, adjusting to the increasing gape and allowing the articular (the small bone in the mandible that articulates with the quadrate) to keep rotating as the gape increased. If these bones had had fixed positions in the skull, the necessary gapes would not have been possible (Gebauer 2007). The jaw-closing muscles must have been very powerful, and the temporal fenestrae, or open spaces in the back of the skull, indicate the need for room for the huge mass of those muscles when contracted.

4.30. A scene in the early Pleistocene of southern Spain, with a group of *Homotherium* hunting a young Meridional mammoth. It is possible that proboscideans in their early years, who occasionally separate from their herds, were adequate prey for sabertooths since they were smaller than adults but not as fiercely protected as the babies.

The post-cranial skeleton is again essentially reptilian, but it reveals a far more upright stance than in more primitive, sprawling synapsids such as the pelycosaurs. The normal locomotion of gorgonopsians would have resembled the so-called high walk of modern crocodilians, with the belly high off the ground, the feet pointing forward, and the limbs carried under the trunk, rather than to the sides. However, the posture of the forelimb, and of the humerus in particular, had a greater horizontal component than the hind limb, so that the elbow pointed outward as the limb advanced but was fairly close to the trunk when the limb was carrying the weight. The motion of the hind limb during the walk was more restricted to the vertical, or sagittal, plane, thus resembling the gait of mammals. As in other reptiles, the musculature of the tail, in particular the caudo-femoral muscles, was a very important part of the flexion of the hind limb, so that the tail was not there merely for balance, as has become the case in mammals. The feet were probably plantigrade, but nonetheless the locomotion of gorgons must have been swift and agile, more than that of most of their prey. The reduction in their phalangeal formula compared to the primitive reptilian condition is seen as an adaptation to make their feet more symmetrical, and their contact with the ground more efficient, as in cursorial mammals.

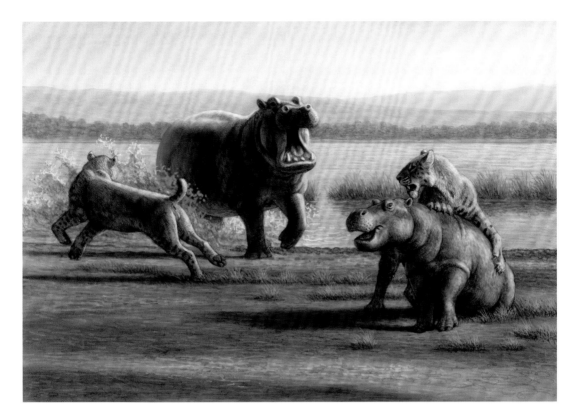

4.31. A scene in the early Pleistocene of southern Spain, with a pair of *Homotherium* attempting to kill a young *Hippopotamus major.*

How would the gorgonopsians hunt? Once the prey was close enough, they would break from cover and swiftly pounce on it, taking advantage of their relatively greater speed (figure 4.34). Using their forelimbs to grab the prey animal, they would bite at any part of its body that they could encompass with the gape of their jaws. Even with the prey still trying to flee, one such bite could cause a large, debilitating blood loss, but as soon as possible the predator would try to bring the prey down and deliver one or more killing bites to the throat or any other vulnerable part of the body.

Growth and Development

There are many different ways in which the fossil record can reveal aspects of the social and family life of sabertooths. One interesting clue concerns the timing of the eruption of the deciduous, or milk, teeth and the permanent ones. Among nimravids and barbourofelids, the pattern is somewhat different from that of true cats or felids, especially in the fact that the permanent canines erupted at a very late stage of development (Bryant 1988). There are fossil skulls of *Eusmilus*, for instance, with quite impressive sabers that on closer inspection turn out to be only the milk canines, with the definitive ones still embedded within the alveolus (figure 4.35). Actually, the milk canines also emerged quite late in the life of the cub, so that although the cub was unable to kill its own prey and

The Social Life of Sabertooths

4.32. A scene in the late Miocene of North America, showing two *Barbourofelis fricki* fighting over a freshly killed protoceratid, *Syntetho-ceras*. Intense competition with conspecifics or with other predators around kills is probably one of the reasons why many sabertooths had proportionally large carnassial teeth, which allowed them to process carcasses very quickly.

thus depended on its mother for a relatively long period, once the milk sabers erupted these were quite large and allowed their owner to hunt for a relatively long time, with the advantage of still having a second chance once the definitive sabers erupted, if the first pair got damaged. In the case of *Barbourofelis*, it is clear that the cubs would have reached almost full adult size and displayed well-worn milk premolars, before the milk sabers even began to erupt, so they would be unable to kill their own prey and would have to remain with their mother or family group until well into their second year. Such long coexistence probably led to situations in which the cubs could contribute their muscular force to help their mother bring down and subdue relatively large prey, a collaboration that might in turn lay the foundation for more extended social ties (Bryant 1990).

In felid sabertooths, the timing of the eruption of the canines was more similar to that in modern cats, but still, if only because of their large size, these teeth took relatively longer to erupt. In Rancho la Brea there are fossil skulls of *Smilodon* individuals in virtually all stages of tooth development and substitution, and many of them display the definitive sabers erupting alongside the smaller milk canines.

But in order to infer the presence of some sort of social groupings in sabertooths, we need to look at other kinds of evidence besides tooth-eruption sequences. Zoologists have long wondered why only modern

4.33. Functional anatomy of the bite in the gorgonopsian *Rubidgea,* with skull, cervical vertebrae, and selected muscles (top); reconstructed life appearance of animal biting at prey (left); and a close-up detail of the reconstructed muzzle of the feeding predator (bottom). The upper canines (1) were the primary killing weapon, while the protruding incisors (2) acted both to stabilize the bitten area (thus protecting the canines from some lateral strains) and to pull chunks of meat off the carcass when feeding. The complex shape of the craniomandibular joint (3) allowed the mandible to remain articulated with the skull even at gapes in excess of 90 degrees. Some ventral muscles of the neck (4), attached to the base, of the skull, contributed to the downward motion necessary to sink the canines into the flesh of the prey. Once the gape was reduced through head depression, the jaw-closing musculature (5) acted to complete the bite.

lions among all the big cats are social, and the most convincing hypothesis to date suggests that the numerical advantage provided by a pride structure when competing with conspecifics for the highest quality territory is the key reason why lions group together (Mosser and Packer 2009). According to this hypothesis, the heterogeneity of savannah habitat would be an important cause of pride formation, because territory quality largely depends on proximity to river confluences, which serve as funnels that force prey into a small area and also hold persistent waterholes and dense vegetation. Lion groups that can defend such optimal territory gain clear advantages over the prides that are forced to the periphery and need to make do with poorer-quality real estate. How does this hypothesis help us infer if a sabertooth species was social or not? All big cats compete for the best territories, but in the case of the lion, being large enough to be

4.34. A scene in the late Permian of Russia, with the gorgonopsian *Inostrancevia* hunting *Scutosaurus*.

the dominant cat in the ecosystem and living in a mosaic environment with open, high-visibility sections are factors that turn competition with other members of its species into a critical factor (visibility increases the likelihood of aggressive encounters). Leopards, for instance, have to worry as much or more about encounters with larger predators than with fellow leopards, and they must lead a more discreet life, hiding in the woods for much of the time. This leaves the leopard with little opportunity to engage in the sort of army warfare that lion prides wage against each other. Thus, if a given species of sabertooth was big enough and lived in a mosaic environment broadly comparable to the African savannah, forming groups would be an advantageous survival strategy. This reasoning suggests that sabertooths like *Amphimachairodus*, *Homotherium*, and *Smilodon*, which were among the biggest cats in their environments and lived in mosaic environments with extensive open sections, could have benefited from becoming social (of course, this doesn't mean that they were necessarily social). Smaller sabertooths, and those that more obviously preferred closed habitats, would in all likelihood be solitary (figure 4.36).

Still other lines of reasoning have been followed to try to determine if some sabertooths might have been social. One interesting study compared the relative abundance of different carnivore species as fossils in Rancho la Brea with the responses of modern carnivores to playbacks of herbivore distress calls in the African savannah (Carbone et al. 2009). Among the African carnivores, it is the dominant, group-living species

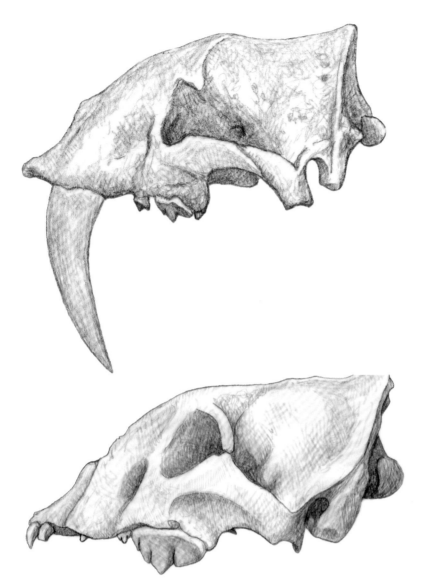

4.35. Sabertooth skulls with milk canines. Top: a juvenile of the nimravid *Eusmilus sicarius,* with fully erupted milk canines and permanent cheek teeth. Bottom: a juvenile of the barbourofelid *Barbourofelis morrisi,* with the milk canine tip just erupting and permanent teeth fully erupted.

like lions and spotted hyenas that show up more often at the playbacks (which act as simulated kill sites); after all, they are confident of their own chances of dominating such a competitive situation. It seems logical, therefore, to infer that in La Brea it would also be the large, social species that would dominate the competitive situation around the trapped herbivores, and since *Smilodon* and the dire wolf *Canis dirus* are by far the most frequently preserved species, the authors find it likely that they were social, too.

One feature of the La Brea sample of *Smilodon* that has been seen as evidence of group behavior is the presence of healed injuries in fossil bones of the sabertooth, since allegedly the injured individuals would have died before their fractures healed unless they were allowed to eat at the kills of other group members (Heald 1989). This idea has been

rebutted by S. MacCall and coauthors (2003), who provide veterinary data about the timing of bone healing in injured felines and about their resistance to starvation, suggesting that the healed injuries found at La Brea are compatible with solitary behavior. According to these authors, such injuries can heal naturally fast enough to allow the animal to survive until it is capable of hunting or at least of opportunistic scavenging. Also, since felids can only get a fraction of their water requirements from their kills, dehydration occurs more rapidly than starvation, and any injured cat must be mobile enough to walk to water during the healing period, a mobility that would also allow it to scavenge opportunistically. Furthermore, the authors argue that the frequency of injuries is comparatively low at the site, suggesting that animals with really severe fractures usually died before healing and would not have been able to get to the asphalt seeps for scavenging. The question of sociability of *Smilodon* from Rancho La Brea remains an open one.

Size Variation and Sexual Dimorphism

Within modern big cat populations, there are size differences not only between cubs and mature animals, but also among adults. Adult males tend to be considerably larger than the females, a phenomenon known as sexual dimorphism (Turner 1984). In order to find out if a fossil species of carnivore was sexually dimorphic, it is necessary to determine if there was a large variation in size and if the sizes of individuals were grouped in a bimodal pattern, meaning that most specimens tend to fall into one of two size groups, rather than there being a continuum of size variation. A large sample is also necessary, and unfortunately fossil sites very rarely yield fossils of many individuals of any carnivore species. So it is always possible that a small sample is giving us a biased picture of the pattern of size variation of the population.

Samples of sabertooth cats generally show a considerable variation of adult sizes, but it is not clear if a sharp dimorphism is implied. Studies of the Rancho la Brea sample of *Smilodon fatalis* have pointed to a very moderate dimorphism, smaller than that of modern lions and possibly comparable to that of modern wolves (Van Valkenburgh and Sacco 2002). A recent study (Meachen-Samuels and Binder 2010) has centered on the percent of pulp cavity closure in the lower canines as a way to improve the accuracy of age estimates, which are an important factor for assigning specimens to sex groups. The results agree with previous studies showing small to nonexistent dimorphism, while the large pantherin cat from the same fossil site, *Panthera atrox*, does show considerable dimorphism, at least as large as that of modern lions.

What are the implications of these results in terms of the behavior of *Smilodon*? Among living carnivores, a high degree of dimorphism is always associated with a high level of competition among the males for access to the females (Gittleman and Van Valkenburgh 1997). In modern pantherin cats, such competition is expressed in two different ways:

4.36. Reconstructed life appearance of a *Smilodon* family group. Several lines of evidence suggest that *Smilodon fatalis* may have benefited from some sort of social grouping. But the very moderate degree of sexual dimorphism in the species suggests that it would not have a harem system like that of modern lions. More likely is a pattern more similar to the extended family groups of canids, with helpers from previous years' litters grouping around a monogamous alpha pair.

among solitary cats like tigers, leopards, and jaguars, males defend large territories, each overlapping the territories of several females, from rival males. In the lion, the only social cat, a male or a coalition of males defend a large territory and a group of resident females from rival males or coalitions. In a way, the harem-like social system of lions can be seen as a derivation of the simpler system of their solitary relatives. A small or absent dimorphism in the sabertooth cat suggests reduced competition, and the authors of the studies agree on the possibility that *Smilodon* was monogamous like canids rather than polygamous like extant cats, a condition that could imply the presence of pair bonding as in jackals or pack life as in wolves.

Batallones-1, one of the few fossil sites to have yielded a significant number of individuals of two Miocene sabertooth species, also allows the study of size variation and possible sexual dimorphism (Antón et al. 2004b; Salesa et al. 2006). The sample from this site suggests that *Machairodus aphanistus* was a dimorphic species, comparable to leopards and lions among modern cats, while *Promegantereon ogygia* shows much less dimorphism. It is possible to infer that males of *Machairodus* would be highly intolerant of the presence of other males, while a greater tolerance can be hypothesized for *Promegantereon*. The greater abundance of *Promegantereon* individuals at the Batallones-1 natural trap could be related to such tolerance. Most individuals recorded were young adults, whose presence was apparently tolerated in parental home ranges—as seems to be the case among modern jaguars, which are both less sexually dimorphic and more tolerant than lions and leopards (Rabinowitz and Nottingham 1986). The greater abundance of *Promegantereon* compared to *Machairodus* may also reflect a greater abundance of the living animals in the ecosystem, to be expected given *Promegantereon*'s smaller body size.

Why Become a Sabertooth?

After reviewing the adaptations of sabertooths, and of sabertooth cats in particular, we have the impression that they were pretty sophisticated and efficient predators, but obviously modern cats are, too. Considering that the anatomical model of the modern or conical-toothed cats is very similar to that of the earliest members of the Felidae, and that it is obviously a model that has worked perfectly well for the last 25 million years, the question inevitably arises of why sabertooth adaptations evolved in the first place. We have seen that the sample of *Machairodus* fossils from Batallones includes a high proportion of individuals with canines that broke in life. Why evolve long, flattened canines that break so easily when they are used against living and kicking prey animals? *Machairodus* was the first member of the cat family to attain the size of a modern tiger or lion, and the obvious reason for it to become so big was to take advantage of the availability of large prey, precisely the kind of prey that could thrash violently when attacked and break the predator's precious sabers—and the same kind of prey that today's conical-toothed cats dispatch successfully with their robust canines. Later members of the machairodontine family would evolve sophisticated adaptations in their teeth, skulls, and post-cranial skeletons that allowed them to kill with less risk of breaking their sabers, but *Machairodus* was a few steps behind in those developments—yet it was the founding father of the homotherin lineage. So if we want to know why the sabertooth adaptations were selected among machairodontines, we must look at precisely that early stage. After all, if the adaptations of *Machairodus* had not given it an edge over competing predators, there would have not been any later stages of homotherin evolution to refine the model.

The case of the other Batallones sabertooth, the leopard-sized *Promeganteron*, is comparable. Its sabers were not as long and flattened as those of *Machairodus* and did not get broken nearly so often. But once again, the animal was clearly developing sabertooth adaptations, and the study of its anatomy provides the clues as to why this model was so successful. Leopard-like as it was in its general proportions, *Promegantereon* differed from modern big cats in having a longer and more powerful neck, a more developed mastoid process, and stronger forelimbs with greater grasping abilities (Salesa et al. 2005, 2006, 2010b). Used in conjunction with the modest sabers, these adaptations meant one thing: *Promagantereon* dispatched its prey far more quickly and safely than any leopard or cougar could. Holding the prey motionless with its strong forelimbs, the cat prevented injury to itself, but the fact that its type of teeth caused massive blood loss rather than suffocation implied that death was a lot quicker. It has been pointed that the bite of modern cats is stronger than that of sabertooths, but the enormous power of the biting muscles of modern cats is needed not so much for crushing or piercing the throat of prey, but for holding their iron grip for many minutes (McHenry et al. 2007). If they let go for even a couple of seconds to readjust the bite, the prey can take a breath and the whole process has to begin again. Sabertooths, even relatively primitive ones like those from Batallones, were in a better position: once they cut a major blood vessel, the weakening and death of the prey was almost a matter of seconds.

The Batallones fossils have helped us to understand why the sabertooth adaptations were first selected, especially among sabertoothed felids, and they have also shown us precisely how the adaptations were selected. As we saw in chapter 3, the craniodental anatomy of *Machairodus aphanistus*, as revealed by the Batallones sample, was a prime example of mosaic evolution, with some features surprisingly derived and others notoriously primitive in their condition. But the relevant detail in evolutionary terms is that the most derived features of *Machairodus aphanistus* were, by far, its upper canine teeth (Antón et al. 2004b). Huge; flattened; and with coarse, serrated margins, the sabers of *M. aphanistus* were so advanced that they would have been at home in any Pleistocene fossil site. However, they were part of a skull that was almost as primitive as that of the ancestral *Pseudaelurus quadridentatus*, and functionally more akin to the skull of a lion than to that of Pleistocene sabertooths like *Smilodon* or *Homotherium*. So, in evolutionary terms, we know the answer to the riddle of what came first, the saber or the sabertooth? At least in this particular group, the saber made the sabertooth.

In the Vallesian ecosystems, with their balanced vegetational cover and steady supply of large mammalian herbivores, *Promegantereon* and *Machairodus* reaped the benefits of their winning design. Later members of their lineages, such as *Megantereon* or *Homotherium*, diverged more spectacularly from the original model of ancestral cats, but in a way they were merely capitalizing on the successful initial moves of the machairodontine pioneers.

In short, the answer to the question "Why become a sabertooth?" is "To kill large prey in a quicker and safer manner." It is no coincidence that the ancestors of our big cats remained in the shade of the sabertooths for most of the Neogene; there was simply no contest in terms of the efficiency with which the machairodontines dealt with large prey. But success can be lethal. We of all species should be aware of that truth.

Extinctions

5

THE FACT THAT NO SABERTOOTHS HAVE SURVIVED TO THE PRESENT day has often led to interpretations that see them as somewhat inferior, slow, or even maladaptive creatures that were left behind in the evolutionary race, replaced by the more fit "normal" big cats. The paleontological literature from the early part of the twentieth century abounds in negative judgments of the biological prowess of sabertooths, from those authors who doubted that the animals could hunt live prey at all to those who questioned even their ability to eat efficiently from carcasses. One almost suspects that the authors of some of this literature felt that the sabertooths "deserved" to become extinct. One good example is this quotation from a paper by the American paleontologist J. Hough: "It should be remembered, also, that by the middle Pleistocene the true cats were making their competition felt. *Smilodon* was too large for effective ambush in trees, too stupid for the type of jungle stalking characteristic of the lion or tiger, and too slow to run down its victims. These modes of attack were being developed rapidly among the true Felidae: *Felis atrox*, the lynx and the jaguar – remains of all of which are found at Rancho la Brea" (1950:135). After such a characterization of *Smilodon*, one wonders how it could manage to coexist with the feline cats of the American Pleistocene for hundreds of thousands of years. Why did it not go extinct sooner?

Other scientists, while assuming that sabertooths were good enough at killing their prey, still thought that they were outcompeted by modern feline cats because the latter were faster and more agile (Simpson 1941). According to such views, sabertooths were adapted to hunt gigantic prey animals who were thick-skinned and slow, and when these behemoths disappeared at the end of the ice age, the only available prey were swift horses and antelope. In this new world, brute force and big teeth were not enough for a predator to survive, so sabertooths disappeared and feline cats triumphed. But, as we have already seen, although some sabertooth species were probably able to take prey animals that were larger and more ponderous than extant big cats can manage, other sabertooths preyed on relatively lightweight, agile prey, herbivores that are still not only extant but abundant. At any rate, even for the species that could on occasion take very large prey, medium-sized to large ungulates (including a wide range of ruminants and horses) would be the most abundant prey and would make up the bulk of the sabertooths' diet, simply because they were the most available source of meat. Also, as we shall see below, "normal" big cats have by no means been immune to extinction, and not only are there several species of them that shared the fate of sabertooths, but some of the species that are still with us became confined to much smaller ranges, and some went to the brink of extinction at about the same time as the last sabertooths vanished.

Another early explanation of the animals' extinction saw the whole process of the evolution of sabertooth features as the cause of their demise. That theory, which was mentioned in passing in chapter 4, proposed that sabertooth evolution was part of an irreversible trend, so that with each generation, the sabers tended to grow slightly larger. As the effects of that trend accumulated, the animals supposedly became less and less able to feed, and extinction was thought to be the inevitable outcome. One interesting example of this view was the interpretation by E. D. Cope of what he called the "characters" of the skull of sabertooths:

> As nothing but the characters of the canine teeth distinguished these from the typical felines, it is to these that we must look for the cause of their failure to continue. Prof. Flower's suggestion appears to be a good one, viz: that the length of these teeth became an inconvenience and a hindrance to their possessors. I think there can be no doubt that the huge canines in the Smilodons must have prevented the biting off of flesh from large pieces, so as to greatly interfere with feeding, and to keep the animals in poor condition. The size of the canines is such as to prevent their use as cutting instruments, excepting with the mouth closed, for the latter could not have been opened sufficiently to allow any object to enter it from the front. Even when it opens so far as to allow the mandible to pass behind the apices of the canines, there would appear to be some risk of the latter's becoming caught on the point of one or the other canine, and forced to remain open, causing early starvation. Such may have been the fate of the fine individual of the *S. neogaeus*, Lund, whose skull was found in Brazil by Lund, and which is familiar to us through the figures of De Blainville, etc. (1880:853)

The individual skull mentioned by Cope had its mandible somewhat displaced laterally, but that displacement in all likelihood occurred after death and had nothing to do with the actual cause of the animal's demise.

Concerning the adaptations of sabertooths, we have seen in previous chapters that they were if anything more sophisticated than those of their non-sabertoothed counterparts. But it is especially interesting that Cope would imagine the advanced sabertooth specializations as something that kept their owners "in poor condition." This would imply that generations of animals during millions of years would breed new generations of equally underfit individuals to drag their miserable existence out a little longer, until the arrival of extinction. Of course such a thing could never happen because, in the natural world, any individual – not to mention a species – which was less than fit would quickly be eliminated without descendants.

Although the idea of species evolving along the path of "suicidal" fixed trends is not realistic, it is nonetheless true that evolution is not an infinitely flexible process of adaptation, and it does indeed have constrains. If we look at the evolution of carnivore dentition in general, we see that the primitive condition (in the order Carnivora and in mammals in general) is to retain a generalized tooth row with a relatively complete set of premolars and molars, as exemplified to some degree by modern dogs. But, as we have seen, carnivores that specialize in a more strictly meat-based diet tend to enlarge their blade-like carnassial teeth and reduce the rest of the post-canine dentition, as adaptations to a more efficient processing of meat. In evolutionary terms, the condition of the hypercarnivores is more derived than that of their relatives with broader diets. In addition, it is an observed fact that rarely, if ever, has a lineage that had evolved into a hypercarnivore morphology reversed direction and gone back to a more generalized one (Van Valkenburgh 2007). In this sense, specialization does operate as an irreversible trend. But although each species, be it specialized or generalized, must be perfectly fit and adapted to its niche if it has to survive for even another generation, it is also true that the more specialized a species becomes, the less flexible it will be in the face of environmental changes. This, as we will see below, does seem to have been a factor in the extinctions of many fossil hypercarnivores, including the sabertooths.

One important point to bear in mind is that there has not been a single extinction episode of sabertooths; rather, there have been many such episodes. Gorgonopsian sabertooths were wiped off the face of the planet during the mass extinction at the end of the Permian, and their end is paradoxically the least problematic: it was simply part of a catastrophic die-off, and if we could figure out what killed over 90 percent of all life forms on earth, then we would also solve the mystery of their disappearance.

The extinctions of the more recent, mammalian sabertooths are more complex because although some of them coincide with larger extinction events, others apparently don't. Sabertoothed creodonts disappeared on their own in the late Eocene, an extinction that doesn't normally cause much of a headache to paleontologists, in part because the animals' fossil record is so modest anyway.

Nimravid sabertooths were a much more varied and successful lineage than the machaeroidines, so it strikes us as more remarkable that after a long evolutionary history, they disappeared from North America in the Oligocene together with their non-sabertoothed close relatives (such as *Nimravus* and *Dinaelurictis*), leaving their habitats devoid of any sabertoothed predator, or of any large catlike carnivore, for millions of years (Bryant 1996b). In North America in particular, it was at least 15 million years between the extinction of the last sabertoothed nimravids, such as *Pogonodon* and *Eusmilus*, and the arrival of the barbourofelids in the late Miocene. In Europe, nimravids apparently vanished earlier than in North America, and it was some 10 million years before the primitive barbourofelids immigrated from Africa. It has been proposed that the nimravids were significantly less agile and fast than many of their prey species, and that such a gap in locomotor adaptations would have become more disadvantageous for the predators as the grasslands became more

5.1. A scene in the early Pleistocene fossil site of Fuente Nueva, in southern Spain, with a band of hominids butchering a young mammoth, possibly a victim of predation by *Homotherium*. Scavenging from the kills of larger carnivores has been a source of protein for our ancestors at least since the advent of the genus *Homo*.

widespread during the Oligocene (Bryant 1996b). But the later success of the immigrant barbourofelids in the American Miocene, when grasslands were even more widespread, shows that relatively open environments could still support robust sabertooth predators with non-cursorial locomotion perfectly well. Nimravid disappearances provide a particularly dramatic example of extinction without replacement.

Barbourofelids had evolved in the Old World, and it took them millions of years to attain a level of sabertooth specialization comparable to that which the derived nimravids had reached millions of years before (Morlo et al. 2004). Then, shortly after reaching the climax of their evolution with the spectacular species *Barbourofelis fricki*, they also disappeared for good. The extinction of the barbourofelids has one potential, partial culprit in the form of the felid sabertooths of the genera *Promegantereon*, *Machairodus*, and *Nimravides* which were spreading successfully throughout the northern hemisphere in the late Miocene and probably came into direct competition with the older group.

The sabertoothed marsupial family Thylacosmilidae has a patchy fossil record in South America starting in the Miocene, and its last representative, *Thylacosmilus*, disappeared in the Pliocene, its place taken by the felid sabertooths that invaded the continent after the appearance of the Isthmus of Panama. A scenario of competitive exclusion is a tempting interpretation of this event, but the evidence does not clearly support it. Instead, the timing of events rather suggests that the marsupial sabertooth was already extinct by the time its placental counterparts arrived on the scene (Marshall and Cifelli 1990).

But the sabertooth extinction event that has most puzzled paleontologists was the demise of genera *Homotherium*, *Megantereon*, and *Smilodon* which had been enormously widespread in Africa, Eurasia, and the Americas in the Pliocene and Pleistocene, and whose place was

5.2. A scene in the Pliocene of South Africa, with *Australopithecus africanus* pursued by *Homotherium*. Although it is unlikely that any sabertooth species took hominids on a regular basis, opportunistic predation must have occurred now and then.

taken by extant big cats such as lions, tigers, leopards, and cheetahs in the Old World, and cougars and jaguars in the New World (Turner and Antón 1998). Sabertooth felids of one kind or another had been continually present in Eurasia, Africa, and North America since the Miocene, but after the end of the Pleistocene only feline big cats remained. The persistence of the latter created the impression of their superiority and was interpreted in the terms of an arms race, with the more agile and fleet predators triumphing in a world where speed, rather than brute force, had become the key to survival. But things are not nearly so simple.

As we have seen, sabertooths, like all top predators, depended for their survival on the availability of suitably sized prey, and they were tolerant of variations in the taxonomic composition of their prey base, of the vegetation supporting those herbivores, and of the climate affecting that vegetation. That is why so many species of sabertooths, especially the large Neogene forms, managed to enjoy almost worldwide distributions. *Homotherium*, for instance, ranged from equatorial to arctic latitudes. But that tolerance was not unlimited, and although sabertooths could thrive under very different circumstances, some combinations of variables could have disastrous effects on them. Such unfavorable conditions concurred several times, leading to the extinctions of many species and whole families of sabertooths. The event at the end of the Pleistocene is just one more example, although it differed from earlier cases in the presence of one important factor—us.

Sabertooths and Humans

We humans have a biased perception of big, wild predators, largely because we are predators, too. Eating meat has been a defining trait of humanity since the earliest members of the genus *Homo* invented stone tools in order to butcher ungulate carcasses in the African savannahs of the Pleistocene, over 2 Ma, and this trait shaped our behavior as surely as it shaped our guts, which happen to be much shorter than those of our more vegetarian relatives, the great apes (figure 5.1). It actually appears that this reduction in gut size gave our organisms an important energetic margin needed to maintain another metabolically costly organ—our big brains. Yet in the African savannahs of the Pliocene and Pleistocene, as in all the natural environments where our early evolution took place, we were not only predators, we also were prey (figure 5.2). Today it is very rare for humans to be hunted and eaten by wild predators, but in the distant past, big carnivores were a much more important part of our existence, and we lived in awe and terror of them, almost like any antelope does. The paleontologist C. K. Brain (1981) actually hypothesized that one particular kind of sabertooths, the members of the genus *Dinofelis*, would have been specialist hominid hunters in the African Pliocene. In his view, these animals would not only hunt the hominids but would also take their bodies back to their lairs to eat at leisure, a habit that would have explained the accumulations of hominid remains in the famous Sterkfontein Valley caves of South Africa (figure 5.3).

Our double existence as predators and prey made us keenly aware of the large carnivores around us, admired as models of hunting behavior, hated as competitors, and feared as our ultimate nemesis. At a time before lions and leopards rose to dominance, sabertooths would have been the top cats on the African "block," featuring prominently in our ancestors' nightmares. Then, for more than a million years, spanning the time of the hominids' expansion out of Africa, sabertooths and modern big cats shared the throne of top predators, making the world an exciting but rather dangerous place for early humans to roam. Finally, with the extinction of the last sabertooth species in North America some 10,000 years ago, the modern big cats were left as the undisputed champions of the terrestrial "man-eating" guild.

In our African cradle, the sabertooths had vanished much earlier, sometime between a million and 500,000 years ago, thus giving an early start to the long reign of the extant lion as "king of the beasts," at least in our imagination. For centuries, human kings have adopted the lion as their favorite symbol and totem animal, but even more significant is the widespread identification that tribal shamans feel with the big cats, which are the ultimate alter ego and totem animal, expected to guide the shamans in their exploration of the world of the spirits. Such identification is clearly expressed in a Paleolithic sculpture that is also one of the earliest known examples of three-dimensional art: the lion-headed

5.3. A scene in the Pliocene of the Sterkfontein Valley, South Africa, showing the sabertooth *Dinofelis* feeding on its *Australopithecus africanus* kill. Black-backed jackals await an opportunity to feed on the scraps.

human figurine from Hohlenstein Stadel, which is thought to be about 32,000 years old. Interpretations of its meaning vary, but the parallelism with images created by extant tribal cultures is striking.

The keen interest of early human artists in big cats raises the obvious question: if anatomically modern humans coexisted with sabertooths in the late Pleistocene, is it not to be expected that they represented such impressive animals? The frustrating fact is that there is no clear example of a sabertooth depiction in prehistoric art. As discussed in chapter 4, the famous (and regrettably lost) felid figurine from Isturitz cave, in France, is best interpreted as a representation of a cave lion.

Admiration of big cats has not prevented humans from pushing them to the brink of extinction. It has been mostly after the advent of pastoralism and agriculture that cats have been actively pursued, as herders set out to defend their livestock and clear the fields around their early settlements of beasts perceived as dangerous to people. But early human hunters may have had a different impact on big cats, especially sabertooths, as competitors for the same resource: ungulate meat. We will see this probable competition in more detail below when we review the various likely causes of sabertooth extinctions.

Causes of
Sabertooth
Extinctions

Competition

As we have seen above, traditional views of the demise of Pleistocene sabertooths saw competition with the allegedly faster modern cats as a direct cause of the sabertooths' extinction, since the modern cats would be more efficient at catching the fleet prey species that have dominated the world after the extinction of the ice age megafauna. But this scenario is hardly accurate in terms of what the sabertooths could and could not hunt, and in terms of which predatory model was more efficient. We must also bear in mind that feline cats were hit very hard by the late Pleistocene extinctions, so that in North America it was not only the two sabertooth species but also the American lion and cheetah that disappeared, suggesting that it was not so much a matter of the sabertooths being outcompeted, but rather a problem with the available resources for all the larger carnivores, as we shall discuss in the next section.

In any case, although the simplistic notion of the "survival of the fleetest" does not explain sabertooth extinctions, that does not rule out the possibility that competition did have a role to play at least in some sabertooth extinction events. As mentioned above, the extinction of barbourofelids in Europe, in the Miocene, took place millions of years earlier than their extinction in North America, and it is possible that the rising of true machairodontine felids such as *Machairodus* and *Promegantereon* posed a serious problem for the last European populations of the barbourofelid *Albanosmilus*. In the New World, species of *Barbourofelis* also coexisted for some time with machairodonts such as *Nimravides* and *Machairodus*, a coexistence that did not prevent this barbourofelid from evolving into the large and specialized species *B. fricki*. Nonetheless, the arrival in North America of the Eurasian genus *Amphimachairodus* was quickly followed by the extinction of *Barbourofelis*, which may or may not have resulted from competition with that immigrant. In each of these cases, we have a competing immigrant taxon with an apparently high degree of ecological overlap and a rather short period of coexistence, so competition is at least a possible factor influencing extinction.

Changes in Prey Availability

The extinction of the North American megafauna, including such giants as the mammoths, mastodons, and ground sloths, is no doubt connected with the disappearance of sabertooths, not so much because the giants were their main prey species (which almost certainly wasn't the case) but because their demise was part of the same major extinction event. Other herbivores that disappeared at the same time included all the New World horses of the genera *Equus* and *Hippidion*, several camelids, antilocaprids, and giant bison species, plus the South American native ungulates such as the camel-like *Macrauchenia* and the hippo-like *Toxodon*. All these species had been a huge biomass of herbivore prey for the big cats, and it is only to be expected that their disappearance had consequences for the predators. Of course, large carnivores are tolerant of different compositions of their prey base as long as prey are abundant and accessible, and the fact that *Smilodon* lived on such different sets of prey species in North and South America is testimony to this. But the extinctions of so many species of herbivores do not occur without a cause, and they point at least in part to an environmental crisis that affected the patterns of vegetation. In the case of the extinction at the end of the Permian, the diversity of prey species for the sabertoothed gorgonopsians had been declining steadily for thousands of years. Ultimately, there were hardly any large herbivores left in South Africa, so it is possible that the giant *Rubidgea* was forced to steal small prey from its more agile cousins, like *Aelurognathus*. Such a decrease in prey diversity was in fact part of the overall environmental deterioration that ended in the catastrophic mass extinction.

Environmental Change

As we saw in chapter 2, vegetational cover is essential for big cats for several reasons: it provides cover that allows them to stalk their prey, and it also minimizes competition by allowing cats to hide or find refuge in trees from the attacks of competitors (Seidensticker 1976). In contrast, large expanses of grasses make the presence of herds of grazers possible, which produces huge concentrations of ungulate meat for the cats. Consequently, big cats thrive best in a mosaic of open areas and more closed woods, and such environments in the past may have favored the coexistence of many species of predators (Antón et al. 2005). Dramatic, global fluctuations in the vegetational cover have taken place many times as a consequence of climatic oscillations, especially during the Pleistocene but also at different stages of the Cenozoic. Such fluctuations would affect the carnivores in many ways, for instance by causing an exaggerated level of competition among species and individuals when optimal, mosaic environments shrank in favor of closed forest or arid areas, neither of which favors the existence of rich guilds of large carnivores.

Throughout the Cenozoic, the gradual elevation of the Andes Mountains in South America blocked moisture-laden winds from the Pacific Ocean and created an ever more severe rain shadow in the continent, which in turn resulted in ever more arid terrestrial environments. The decline in the variety of carnivorous marsupials during the Cenozoic may have been related to the decrease of vegetational cover, so important for the predators to lay their ambushes. The sabertoothed marsupial *Thylacosmilus*, with its robust body proportions, was clearly an ambush predator, and its survival would depend on the presence of gallery woodlands, a type of habitat that probably shrank through the Pliocene period. In fact, *Thylacosmilus* went extinct during the most dramatic faunal turnover in the whole Pliocene and Pleistocene of South America, which took place in the Cahapadmalalian-Uquian transition and is attributed to environmental changes more than to the influence of mammal species emigrating from North America (Pascual et al. 1996).

The environmental deterioration at the end of the Permian included aridification and a dramatic loss of vegetational cover, so this factor was also present in the extinction of the gorgonopsian sabertooths.

Human Influence

The adverse effects of climatic oscillations may have been made worse in the Pleistocene by the entrance of humans into the large carnivore guild. As species of our own genus *Homo* became larger and better equipped technologically for aggressive scavenging and the hunting of large prey, their pressure on large carnivores would increase. This influence may have been felt first in Africa, where *Homo ergaster* became widespread as early as 1.6 Ma, with the sabertooth cats *Megantereon* and *Homotherium* becoming extinct shortly after (figure 5.4). The genus *Dinofelis* appears to have survived for several hundred thousand years afterward (Werdelin and Lewis 2001), a fact that may be related to the animals' lesser degree of machairodont specialization and thus potentially greater ecological flexibility.

In Europe the sabertooths lasted much longer, but *Megantereon* went extinct around 1 Ma, after the arrival of *Homo antecessor*. *Megantereon* had shared the environments of the hominid migrants since their earliest expansion out of Africa, and there are good reasons to believe that it suffered some degree of kleptoparasitism (a term meaning that one species systematically steals prey from another, to the detriment of the latter) from our early relatives. In fact, some specialists have hypothesized that the early species of our genus *Homo* developed a real dependence on the

sabertooths as providers of half-eaten carcasses, which meant a crucial protein intake especially during the leaner months of the year (Arribas and Palmqvist 1999). Proponents of that scenario suggested that the first expansion of our early relatives out of Africa was largely facilitated by the presence of the sabertooth *Megantereon* as a fellow traveler.

Such hypotheses are partly based on the idea that sabertooths, with their specialized dentitions, were unable to deflesh their prey as thoroughly as other carnivores do, leaving a considerable amount of flesh on the carcasses, and that their blade-like carnassials would render them unable to crack bones, so that all the nutrients in the bones would remain there for the taking. But it is not clear that sabertooths were actually significantly better providers of partly eaten carcasses than feline big cats are, because the latter have dentitions almost as unsuited as those of sabertooths for processing bone, and once they deflesh a carcass most of the in-bone nutrients are left for scavengers, as would have been the case with sabertooths. Also, the procumbent incisors of sabertooths were probably very efficient tools for use in defleshing bone. The paleontologist C. Marean (1989) studied the fossils from Friesenhahn Cave, a probable den of *Homotherium* in Texas, and found that many bones of young

5.4. A scene in the early Pleistocene of Koobi Fora, Kenya, with a band of *Homo ergaster* trying to evict *Megantereon* from its waterbuck kill. With a larger body size than earlier hominid species and more refined stone technology, *Homo ergaster* probably turned from opportunistic scavenging to true kleptoparisitism.

proboscideans had marks that fitted perfectly with the defleshing action of the incisors of the scimitar-tooth cat. Additionally, the incisors themselves often showed strong wear, indicating that they had been heavily used and may have often been in contact with bone. Such evidence indicates that the supposed inability of sabertooths to thoroughly consume the flesh of their kills is open to question.

In the case of *Megantereon*, the fact that it was solitary and not very big probably did make it a better target for aggressive human scavenging than larger and probably group-living predators like *Homotherium*. Hominids are indeed associated with sabertooths in the fossil sites documenting their early occupation of Eurasia, such as Dmanisi in Georgia. The abundance of stone tools around some of the animal carcasses at Dmanisi makes it easy to imagine the quick arrival of the hominids to this area where many of the kills of *Megantereon* probably took place, and in fact one of the skulls of *Homo georgicus* from that site has an injury that is best interpreted as a wound caused by the upper canine of the sabertooth, a testimony to the conflicts arising around the uneasy sharing of carcasses (Gabunia et al. 2000).

Homotherium survived longer than *Megantereon* in Europe, but after the arrival of the larger and technologically better equipped *Homo heidelbergensis*, it also seems to disappear. After an absence in the fossil record of 400,000 years, *Homotherium* is detected as a single fossil mandible 28,000 years old in the North Sea, a record that may mean either the persistence of a very small population in the shadow of the ever-present lions, or a re-immigration from North America. In any case, nothing contradicts the picture of an overwhelming dominion of the lion in the big-cat niche in Eurasia during the last 400,000 years.

In the Americas, the sabertooths survive for much longer, but the arrival of modern *Homo sapiens* on the continent marks the disappearance of *Smilodon* and *Homotherium*.

The Vulnerability of Specialists

The above narration of the extinction or extinctions of *Homotherium* in Europe reveals a very interesting combination of several possible contributing factors. The arrival of the lion from Africa in the middle Pleistocene introduced a factor of competition, which may have affected *Homotherium* to some degree, and the later arrival of *Homo heidelbergensis* clearly contributed to what had become a dangerously crowded guild of large predators.

It is impossible to know if these competition factors by themselves would have pushed the sabertooth to extinction, but the compelling evidence for strong climatic oscillations during the whole Pleistocene points to another ingredient that was making life more difficult for all those predators. As we have seen, the optimum environment for the presence of a rich carnivore guild is one that includes a mosaic of open and closed vegetation, where predators can ambush their prey and avoid too many dangerous encounters with competitors. During both the warm and the cold climatic extremes of the Pleistocene, such optimum environments would contract while either closed forests (during the warm interglacials) or steppes (during cold glacials) expanded, and during such episodes, competition between carnivores would be exacerbated. The continued abundance of lion fossils through all climatic extremes in the European Pleistocene record shows that the lion had a clear edge over the sabertooth, and that edge was most likely caused by its greater adaptability to changes in its prey base. African lions today can subsist on a surprising variety of ungulate species, turning to small, unlikely prey when larger herbivores are seasonally absent. But sabertooths' specialized anatomy would have left them less able to exploit small prey, and their adaptations to kill large animals so efficiently turned against them during leaner times.

In the Americas, Pleistocene climatic oscillations were also severe, but the arrival of humans occurred much later than in Europe, and it was followed by a catastrophic extinction of the megafauna. So there is reason to suspect that although the inflated large-carnivore guild of the American Pleistocene could cope with many cycles of climate oscillation, the combination of environmental change and the human presence was just too much for the sabertooths.

Evidence thus suggests that a combination of climatic-induced environmental crises, unpredictable distribution of resources, and the concomitant increase in intraguild competition all had an important role as possible causes for Pleistocene sabertooth extinctions. This makes us suspect that comparable combinations of causes have been at work in previous extinction episodes. The worldwide drying and opening of environments in the late Oligocene may have forced the last nimravid sabertooths to live in low densities over enormous ranges, and during some episodes of dramatically dry climate it is possible that the widespread amphicyonid bear-dogs, with their all-terrain locomotion and wide-spectrum diet, had an edge over the narrow-niche sabertooths. A group of medium-sized bear dogs, the temnocyonines, were the only large carnivores to survive the Oligocene-Miocene boundary faunal turnover in North America, and the new amphicyonid and ursid species arriving from the Old World in the early Miocene apparently entered a continent almost devoid of large predators (Hunt 2002). Whatever combination of causes finished off the nimravid sabertooths, it also took almost every other mammalian predator on the continent, including the dog-like hyaenodontind creodonts.

Ironically, it is possible that millions of years later, the large felid sabertooths such as *Machairodus*, with their devastatingly efficient predation methods, had an advantage during the more stable, favorable conditions of the Eurasian Vallesian epoch, helping to push the last amphicyonids to extinction. One is even tempted to see a pattern here, in which specialist predators thrive during periods when resources are plentiful and predictable, and generalists have the advantage during times of crisis.

If we could release some of the Pleistocene sabertooths in suitable modern ecosystems, it is possible that they might hold their ground against competing predators and thrive again on a diet of buffalo, bison, and other large ungulates. Feral horses, for instance, prosper in North America, showing that their extinction in that continent was not due to some radical change that made the environments there forever unsuitable for horses; sabertooths might be a similar case.

Extinction is often just a consequence of many things going wrong at the same time for a species. Or, to put it in other words, we see contingency in action, rather than a predestined fate. Once the crisis is past, any survivors of a decimated species can repopulate its previous domains, and we might never suspect how close it was to extinction. We do know that the cheetah was on the brink some 10,000 years ago, because the extreme genetic uniformity of all living members of the species is evidence of a severe population bottleneck at that time (O'Brien et al. 1987). It is tempting to see the cheetah as the opposite of the sabertooths, fast and agile where they were slow and strong. But the animals have more in common than one might realize at first. They share their vulnerability as extreme specialists, and we may be very wrong to think that the cheetah is better adapted to modern environments than the sabertooths were. In fact, it may be mere luck that the cheetah is still with us instead of following the fate of sabertooths – or the fate of its cousin, the American cheetah, for that matter. There are interesting although sobering lessons in this for us, if we are to assume our proper role as stewards of biodiversity on this planet.

Bibliography

Adolfssen, J. S., and P. Christiansen. 2007. Osteology and ecology of *Megantereon cultridens* SE311 (Mammalia; Felidae; Machairodontinae), a sabrecat from the late Pliocene–early Pleistocene of Senèze, France. *Zoological Journal of the Linnean Society,* 151: 833–884.

Agustí, J., and M. Antón. 2002. *Mammoths, Sabertooths and Hominids: 65 Million Years of Mammalian Evolution in Europe.* Columbia University Press, New York.

Akersten, W. A. 1985. Canine function in *Smilodon* (Mammalia, Felidae, Machairodontinae). *Los Angeles County Museum Contributions in Science,* 356: 1–22.

Alf, M. 1959. Mammal footprints from the Avawatz Formation, California. *Bulletin of Southern California Academy of Science,* 58: 1–7.

Alf, M. 1966. Mammal trackways from the Barstow formation, California. *Bulletin of Southern California Academy of Science,* 65: 258–264.

Antón, M., and A. Galobart. 1999. Neck function and predatory behavior in the scimitar toothed cat *Homotherium latidens* (Owen). *Journal of Vertebrate Paleontology,* 19: 771–784.

Antón, M., A. Galobart, and A. Turner. 2005. Co-existence of scimitar-toothed cats, lions and hominins in the European Pleistocene: implications of the post-cranial anatomy of *Homotherium latidens* (Owen) for comparative palaeoecology. *Quaternary Science Reviews,* 24: 1287–1301.

Antón, M., R. García-Perea, and A. Turner. 1998. Reconstructed facial appearance of the sabretoothed felid *Smilodon. Zoological Journal of the Linnean Society,* 124: 369–386.

Antón, M., G. López, and R. Santamaría. 2004a. Carnivore trackways from the Miocene site of Salinas de Añana (Álava, Spain). *Ichnos,* 11: 371–384.

Antón, M., and J. Morales. 2000. Inferencias paleoecológicas de la asociación de carnívoros del yacimiento de Cerro Batallones. In J. Morales, M. Nieto, L. Amezua, S. Fraile, E. Gómez, E. Herráez, P. Peláez-Campomanes, M. J. Salesa, I. M. Sánchez, and D. Soria (eds.), *Patrimonio Paleontológico de la Comunidad de Madrid,* 190–201. Arqueología, Paleontología y Etnografía, 6. Servicio de Publicaciones de la Comunidad de Madrid, Madrid.

Antón, M., M. J. Salesa, A. Galobart, J. F. Pastor, and A. Turner. 2009. Soft tissue reconstruction of *Homotherium latidens* (Mammalia, Carnivora, Felidae): Implications for the possibility of representation in Palaeolithic art. *Geobios,* 42: 541–551.

Antón, M., M. J. Salesa, J. Morales, and A. Turner. 2004b. First known complete skulls of the scimitar-toothed cat *Machairodus aphanistus* (Felidae, Carnivora) from the Spanish late Miocene site of Cerro Batallones-1. *Journal of Vertebrate Paleontology,* 24: 957–969.

Antón, M., M. J. Salesa, J. F. Pastor, I. M. Sanchez, S. Fraile, and J. Morales. 2004c. Implications of the mastoid anatomy of larger extant felids for the evolution and predatory behaviour of sabretoothed cats (Mammalia, Carnivora, Felidae). *Zoological Journal of the Linnean Society,* 140: 207–221.

Antón, M., and L. Werdelin. 1998. Too well restored? The case of the *Megantereon* skull from Senèze. *Lethaia,* 31: 158–160.

Anyonge, W. 1993. Body mass in large extant and extinct carnivores. *Journal of Zoology,* 231: 339–350.

Arambourg, C. 1947. Contribution à l'étude géologique et paléontologique du bassin du Lac Rodolphe et de la basse vallée de l'Omo, deuxième partie. In C. Arambourg (ed.), *Mission Scientifique de l'Omo, Paléontologie,* 1: 231–562. Editions du Museum, Paris.

Arambourg, C. 1970. Les vertébrés du Pléistocène de l'Afrique du nord. *Archives du Muséum national d'Histoire naturelle, Paris,* 7th ser., 10: 1–127.

Argant, A. 2004. Les carnivores du gisement Pliocène final de Saint-Vallier (Drôme, France). Supplement 1, Le gisement pliocène final de Saint-Vallier (Drôme, France). *Geobios 37:* S133–S182.

Argot, C. 2004. Functional-adaptive features and palaeobiologic implications of the postcranial skeleton of the late Miocene sabretooth borhyaenoid *Thylacosmilus atrox* (Metatheria). *Alcheringa,* 28 (1): 229–266.

Arribas, A., and P. Palmqvist. 1999. On the ecological connection between sabre-tooths and hominids: faunal dispersal events in the lower Pleistocene and a review of the evidence for the first human arrival in Europe. *Journal of Archaeological Science,* 26: 571–585.

Bailey, T. N. 1993. *The African Leopard: Ecology and Behavior of a Solitary Felid.* Columbia University Press, New York.

Bakker, R. T. 1998. Brontosaur killers: late Jurassic allosaurids as sabretooth cat analogues. *Gaia,* 15: 145–158.

Ballesio, R. 1963. Monographie d'un Machairodus du gisement villafranchien de Senèze: *Homotherium crenatidens* Fabrini. *Travaux du Laboratoire de Geologie, Lyon,* 9: 1–127.

Barnett, R., I. Barnes, M. J. Phillips, L. D. Martin, C. R. Harington, J. A. Leonard, and A. Cooper. 2005. Evolution of the extinct sabretooths and the American cheetahlike cat. *Current Biology,* 15: R589–R590.

Barone, R. 2010. *Anatomie Comparée des Mammifères Domestiques.* Vols. 1 and 2. Éditions Vigot, Paris.

Baskin, J. A. 2005. Carnivora from the Late Miocene Love Bone Bed local fauna of Florida. *Bulletin of the Florida Museum of Natural History,* 45: 413–434.

Battail, B., and M. V. Surkov. 2000. Mammal-like reptiles from Russia. In M. J. Benton, M. A. Shishkin, D. M. Unwin, and E. N. Kurochkin (eds.), *The Age of Dinosaurs in Russia and*

Mongolia, 86–119. Cambridge University Press, Cambridge.

Beaumont, G. de. 1975. Recherches sur les Félidés (Mammifères, Carnivores) du Pliocène inférieur des sables à Dinotherium des environs d'Eppelsheim (Rheinhessen). Archives des Sciences Physiques et Naturelles, Genève, 28: 369–405.

Bernor, R. L., H. Tobien, L. C. Hayek, and H. W. Mittmann. 1997. Hippotherium primigenium (Equidae, Mammalia) from the late Miocene of Höwenegg (Hegau, Germany). Andrias, 10: 1–230.

Berta, A. 1987. The sabercat Smilodon gracilis from Florida and a discussion of its relationships (Mammalia, Felidae, Smilodontini). Bulletin of the Florida State Museum, Biological Series, 31: 1–63.

Blainville, H. M. D. de. 1841. Ostéographie ou description iconographique comparée du squelette et du système dentaire des mammifères récents et fossiles pour servir de base à la zoologie et à la géologie. Vol. 2, Carnassiers. Baillière, Paris.

Boaz, N. T., R. L. Ciochon, Q. Xu, and J. Liu. 2000. Large mammalian carnivores as a taphonomic factor in bone accumulation at Zhoukoudian. Supplement. Acta Anthropologica Sinica, 19: 224–234.

Bohlin, B. 1940. Food habit of the machairodonts, with special regard to Smilodon. Bulletin of the Geological Institute of Uppsala, 28: 157–174.

Bohlin, B. 1947. The sabre-toothed tigers once more. Bulletin of the Geological Institute of Uppsala, 32: 11–20.

Boule, M. 1901. Révision des espèces européennes de Machairodus. Bulletin de la Société Géologique de France, 4: 551–573.

Brain, C. K. 1981. The Hunters or the Hunted? An Introduction to African Cave Taphonomy. University of Chicago Press, Chicago.

Bravard, M. A. 1828. Monographie de la Montagne de Perrier, près d'Issoire (Puy-de-Dôme) et de deux espèces fossiles du genre Felis découvertes dans l'une de ses couches d'alluvions. Dufour et Docagne, Amsterdam, and Levrault, Strasbourg.

Broom, R. 1925. On some carnivorous therapsids. Records of the Albany Museum, 3: 309–326.

Broom, R. 1937. On some new Pleistocene mammals from limestone caves of the Transvaal. South African Journal of Science, 33: 750–768.

Broom, R. 1938. On a new family of carnivorous therapsids from the Karroo beds of South Africa. Proceedings of the Zoological Society of London, B108: 527–533.

Bryant, H. N. 1988. Delayed eruption of the deciduous upper canine in the saber-toothed carnivore Barbourofelis lovei (Carnivora Nimravidae). Journal of Vertebrate Paleontology, 8: 295–306.

Bryant, H. N. 1990. Implications of the dental eruption sequence in Barbourofelis (Carnivora, Nimravidae) for the function of upper canines and the duration of parental care in sabretoothed carnivores. Journal of Zoology, 222: 585–590.

Bryant, H. N. 1991. Phylogenetic relationships and systematics of the Nimravidae (Carnivora). Journal of Mammalogy, 72: 56–78.

Bryant, H. N. 1996a. Force generation by the jaw adductor musculature at different gapes in the Pleistocene sabretoothed felid Smilodon. In K. M. Stewart and K. L. Seymour (eds.), Palaeoecology and Palaeoenvironments of Late Cenozoic Mammals, 283–299. University of Toronto Press, Toronto.

Bryant, H. N. 1996b. Nimravidae. In D. R. Prothero and R. J. Emry (eds.), The Terrestrial Eocene-Oligocene Transition in North America, 453–475. Cambridge University Press, Cambridge.

Bryant, H. N., and A. P. Russell. 1992. The role of phylogenetic analysis in the inference of unpreserved attributes of extinct taxa. Philosophical Transactions of the Royal Society of London, B337: 405–418.

Carbone, C., T. Maddox, P. J. Funston, M. G. L. Mills, G. F. Grether, and B. Van Valkenburg. 2009. Parallels between playbacks and Pleistocene tar seeps suggest sociality in an extinct sabretooth cat, Smilodon. Royal Society Biology Letters, 5: 81–85.

Cartelle, C. 1994. Tempo passado: Mamiferos do Pleistoceno em Minas Gerais. Palco, Belo Horizonte, Brasil.

Catuneanu, O., H. Wopfner, P. G. Eriksson, B. Cairncross, B. S. Rubidge, R. M. H. Smith, and P. J. Hancox. 2005. The Karoo basins of south-central Africa. Journal of African Earth Sciences, 43: 211–253.

Chang, H. 1957. On new material of some machairodonts of Pontian age from Shansi. Vertebrata Palasiatica, 1: 193–200.

Christiansen, P., and J. M. Harris. 2005. Body size of Smilodon (Mammalia: Felidae). Journal of Morphology, 266: 369–384.

Colbert, E. H. 1948. The mammal-like reptile Lycaenops. Bulletin of the American Museum of Natural History, 89: 357–404.

Cooke, H. B. S. 1991. Dinofelis barlowi (Mammalia, Carnivora, Felidae) cranial material from Bolt's Farm, collected by the University of California African Expedition. Palaeontologia Africana, 28: 9–21.

Cope, E. D. 1879. Scientific news. American Naturalist 13: 798a–798b.

Cope, E. D. 1880. On the extinct cats of America. American Naturalist, 14: 833–858.

Cope, E. D. 1887. A saber-tooth tiger from the Loup Fork Beds. American Naturalist, 21: 1019–1020.

Cope, E. D. 1893. A Preliminary Report on the Vertebrate Paleontology of the Llano Estacado. Geological Survey of Texas, B. C. Jones, Austin, Texas.

Cox, S. M., and G. T. Jefferson. 1988. The first individual skeleton of Smilodon from Rancho La Brea. Current Research in the Pleistocene, 5: 66–67.

Creel, S., and N. M. Creel. 2002. The African Wild Dog: Behavior, Ecology, and Conservation. Princeton University Press, Princeton.

Croizet, J. B., and A. C. G. Jobert. 1828. Recherches sur les ossements fossiles du departement de Puy-de-Dôme. Delahays, Paris.

Crusafont, M., and E. Aguirre. 1972. Stenailurus, felide nouveau, du Turolien d'Espagne. Annales des Paleontologie (Vertebres), 58: 211–223.

Cuvier, G. 1824. Recherches sur les ossements fossiles, où l'on rétablit les caractères de plusieurs animaux dont les révolutions du globe ont détruit les espèces. Vol 5. Edmund d'Ocagne, Paris.

Dalquest, W. W. 1969. Pliocene carnivores of the Coffee Ranch (type Hemphill) local fauna. Bulletin of the Texas Memorial Museum, 15: 1–44.

Dalquest, W. W. 1983. Mammals of the Coffee Ranch local Hemphillian fauna of Texas (USA). PearceSellards Series, Texas Memorial Museum, 38: 1–41.

Davis, D. D. 1964. The giant panda: a morphological study of evolutionary mechanisms. Fieldiana: Zoology Memoirs, 3: 1–339.

Dawson, M. R., R. K. Stucky, L. Krishtalka, and C. C. Black. 1986. *Machaeroides simpsoni,* new species, oldest known sabertooth creodont (Mammalia), of Lost Cabin Eocene. In K. M. Flanagan and J. A. Lillegraven (eds.), *Vertebrates, Phylogeny, and Philosophy,* 177–182. University of Wyoming Department of Geology and Geophysics, Laramie.

Dechamps, R., and F. Maes. 1985. Essai de reconstitution des climats et des vegetations de la basse vallee de l'Omo au Plio-Pleistocene à l'aide de bois fossiles. In M. Beden (ed.), *L'Environnement des Hominidés au Plio-Pleìstoceàne: colloque international (juin 1981),* 175–221. Fondation Singer-Polignac, Masson, Paris.

Delson, E., M. Faure, C. Guérin, A. Aprile, J. Argant, B. A. B. Blackwell, E. Debard, W. Harcourt-Smith, E. Martin-Suarez, A. Monguillon, F. Parenti, J.-F. Pastre, S. Sen, A. R. Skinner, C. C. Swisher III, and A. M. F. Valli. 2006. Franco-American renewed research at the Late Villafranchian locality of Senèze (Haute-Loire, France). *Courier Forschunginstitut Senckenberg,* 256: 275–290.

Deng, T. 2006. Paleoecological comparison between late Miocene localities of China and Greece based on Hipparion faunas. *Geodiversitas,* 28: 499–516.

Diamond, J. 1986. How great white sharks, sabre-toothed cats and soldiers kill. *Nature,* 322: 773– 774.

Evernden, J. F., D. E. Savage, G. H. Curtis, and G. T. James. 1964. Potassium-argon dates and the Cenozoic mammalian chronology of North America. *American Journal of Science,* 262: 145–198.

Ewer, R. F. 1955. The fossil carnivores of the Transvaal caves: Machairodontinae. *Proceedings of the Zoological Society of London,* 125: 587–615.

Fabrini, E. 1890. *Machairodus (Meganthereon)* del Val d'Arno superiore. *Bolletino de Reale Comitato Geologico d'Italia,* ser. 3, 1: 121–144 and 161–177.

Falconer, H., and P. T. Cautley. 1836. Note on the *Felis cristata,* a new fossil tiger from the Sivalik Hills. *Asiatic Researches,* 19: 135–142.

Ficcarelli, G. 1979. The Villafranchian machairodonts of Tuscany. *Palaeontographia Italica,* 71: 17–26.

Filhol, M. 1872. Note relative à la découverte dans les gisements de phosphate de chaux du Lot d'un Mammifère fossile nouveau. *Bulletin de la Société des Sciences Physiques et Naturelles, Toulouse,* 1: 204–208.

Forasiepi, A., and A. A. Carlini. 2010. A new thylacosmilid (Mammalia, Metatheria, Sparassodonta) from the Miocene of Patagonia. *Zootaxa,* 2552: 55–68.

Gabunia, L., A. Vekua, and D. Lordkipanidze, 2000. The environmental contexts of early human occupation of Georgia (Transcaucasia). *Journal of Human Evolution,* 38: 785–802.

Galobart, A. 2003. Aspectos tafonómicos de los yacimientos del Pleistoceno inferior de Incarcal (Crespiá, NE de la Península Ibérica). *Paleontología y Evolució,* 34: 211–220.

Gaudry, A. 1862 and 1867. *Animaux fossiles et géologie de l'Attique.* 2 vols. F. Savy, Paris.

Gazin, C. L. 1946. *Machaeroides eothen* Matthew, the sabertooth creodont of the Bridger Eocene. *Proceedings of the United States National Museum,* 96: 335–347.

Gebauer, E. 2007. Phylogeny and evolution of the Gorgonopsia with a special reference to the skull and skeleton of GPIT/RE/7113 ("Aelurognathus? Parringtoni"). PhD dissertation, Tübingen University, Tübingen, Germany.

Geraads, D., and E. Gulec. 1997. Relationships of *Barbourofelis piveteaui,* a late Miocene nimravid from central Turkey. *Journal of Vertebrate Palaeontology,* 17: 370–375.

Gervais, P. 1876. *Zoologie et paléontologie françaises.* 2nd ed. Arthus Bertrand, Paris.

Gillette, D. D., and C. E. Ray. 1981. Glyptodons of North America. *Smithsonian Contributions to Paleobiology,* 40: 1–255.

Ginsburg, L. 1961a. La faune des carnivores miocènes de Sansan. *Mémoires du Muséum national d'Histoire naturelle,* 9: 1–190.

Ginsburg, L. 1961b. Plantigradie et digitigradie chez les carnivores fissipèdes. *Mammalia,* 25: 1–21.

Ginsburg, L. 2000. La faune miocène de Sansan et son environnement. *Mémoires du Muséum national d'Histoire naturelle,* 183: 9–10.

Ginsburg, L., J. Morales, and D. Soria. 1981. Nuevos datos sobre los carnívoros de Los Valles de Fuentidueña, Segovia. *Estudios Geológicos,* 37: 383–415.

Gittleman, J. J., and B. Van Valkenburgh. 1997. Sexual dimorphism in the canines and skulls of carnivores: effects of size, phylogeny, and behavioural ecology. *Journal of Zoology,* 242: 97–117.

Goin, F. J. 1997. New clues for understanding Neogene marsupial radiations. In R. F. Kay, R. Cifelli, R. H. Madden, and J. Flynn (eds.), *Vertebrate Paleontology in the Neotropics,* 185–204. Smithsonian Institution Press, Washington.

Goin, F. J., and R. Pascual. 1987. News on the biology and taxonomy of the marsupials Thylacosmilidae (Late Tertiary of Argentina). *Anales de la Academia Nacional de Ciencias Exactas, Físicas y Naturales, Buenos Aires,* 39: 219–246.

Harris, J. M., and G. T. Jefferson (eds.). 1985. *Rancho la Brea: Treasures from the Tar Pits.* Natural History Museum of Los Angeles County, Los Angeles.

Hatcher, J. B. 1895. Discovery, in the Oligocene of South Dakota, of *Eusmilus,* a genus of saber-toothed cat new to North America. *American Naturalist,* 29: 1091–1093.

Heald, F. 1989. Injuries and diseases in *Smilodon californicus. Journal of Vertebrate Paleontology,* 9: 24A.

Hearst, J. M., L. D. Martin, J. P. Babiarz, and V. L. Naples. 2011. Osteology and myology of *Homotherium ischyrus* from Idaho. In V. L. Naples, L. D. Martin, and J. P. Babiarz (eds.), *The Other Saber-Tooths: Scimitar-Tooth Cats of the Western Hemisphere,* 123–183. Johns Hopkins University Press, Baltimore.

Hemmer, H. 1965. Zur nomenklatur und verbreitung des genus *Dinofelis* Zdansky, 1924 (*Therailurus* Piveteau, 1948). *Palaeontologia Africana,* 9: 75–89.

Hemmer, H. 1978. Socialization by intelligence. *Carnivore,* 1: 102–105.

Hildebrand, M. 1959. Motions of the running cheetah and horse. *Journal of Mammalogy,* 40: 481–495.

Hildebrand, M. 1961. Further studies on locomotion of the cheetah. *Journal of Mammalogy,* 42: 84–91.

Hoganson, J. W., E. C. Murphy, and N. F. Forsman. 1998. Lithostratigraphy, paleontology, and biochronology of the Chadron, Brule, and Arikaree Formations in North Dakota. In D. O. Terry Jr., H. E. La Garry, and R. M. Hunt Jr. (eds.), *Depositional Environments, Lithostratigraphy, and Biostratigraphy of the White River and Arikaree Groups (Late Eocene to Early Miocene), North America,* 185–196. Geological Society of America, Boulder, Colorado.

Hough, J. 1950. The habits and adaptation of the Oligocene saber tooth carnivore, *Hoplophoneus*. US Geological Survey Professional Paper 221-H. Government Printing Office, Washington. http://pubs.usgs.gov/pp/0221h/report.pdf.

Hudson, P. E., S. A. Corr, R. C. Payne-Davis, S. N. Clancy, E. Lane, and A. M. Wilson. 2011. Functional anatomy of the cheetah (Acinonyx jubatus) hindlimb. *Journal of Anatomy,* 218: 363–374.

Huene, F. von. 1950. Die Theriodontier des ostafricanischen Ruhuhu-Gebietes in der Tübinger Sammlung. *Neues Jahrbuch Geologie und Paläontologie,* 92: 47–136.

Hunt, R. M. 1987. Evolution of the aeluroid Carnivora: significance of the auditory structure in the nimravid cat, *Dinictis. American Museum Novitates,* 2886: 1–74.

Hunt, R. M. 2002. Intercontinental migration of Neogene Amphicyonids (Mammalia, Carnivora): appearance of the Eurasian bear-dog *Ysengrinia* in North America. *American Museum Novitates,* 3384: 1–53.

Janis, C. 1994. The sabertooth's repeat performances. *Natural History,* 103: 78–83.

Jepsen, G. L. 1933. American eusmiloid sabre tooth cats of the Oligocene epoch. *Proceedings of the American Philosophical Society,* 72: 355–369.

Johnston, C. S. 1937. Tracks from the Pliocene of West Texas. *American Midland Naturalist,* 18: 147–152.

Kaup, J. J. 1832. Vier neue Arten urweltlicher Raubthiere welche im zoologischen Museum zu Darmstadt aufbewart werden. *Archiv für Mineralogie,* 5: 150–158.

Kovatchev, D. 2001. Description d'un squelette complet de *Metailurus* (Felidae, Carnivora, Mammalia) du Miocène supérieur de Bulgarie. *Geologica Balcanica,* 31: 71–88.

Kretzoi, M. 1929. Materialen zur phylogenetischen klassifikation der Aeluroideen. In E. Csiki (ed.), *Comptes Rendus, 10th International Zoological Congress, tenu á Budapest du 4 au 10 septembre 1927,* 1293–1355. Stephaneum, Budapest.

Kretzoi, M. 1938. Die raubtiere von Gombaszög nebst einer Übersicht der Gesamtfauna. *Annales Historico-Naturales Musei Nationalis Hungarici,* 31: 88–157.

Kurtén, B. 1968. *Pleistocene Mammals of Europe.* Weidenfeld and Nicolson, London.

Kurtén, B. 1976. Fossil Carnivora from the late Tertiary of Bled Douarah and Cherichira, Tunisia. *Notes du Service géologique de Tunisie,* 42: 177–214.

Kurtén, B. 1980. *Dance of the Tiger.* Random House, New York.

Kurtén, B., and E. Anderson. 1980. *Pleistocene Mammals of North America.* Columbia University Press, New York.

Leakey, M. G., and J. M. Harris. 2003. Lothagam: its significance and contributions. In M. G. Leakey and J. M. Harris (eds.), *Lothagam: The Dawn of Humanity in Eastern Africa,* 625–660. Columbia University Press, New York.

Leidy, J. 1851. [Untitled article.] *Proceedings of the Academy of Natural Sciences of Philadelphia,* 5: 329–330.

Leidy, J. 1854. The ancient fauna of Nebraska, or a description of remains of extinct Mammalia and Chelonia, from the Mauvaises Terres of Nebraska. *Smithsonian Contributions to Knowledge,* 6: 1–126.

Leidy, J. 1868. Notice of some vertebrate remains from Hardin County, Texas. *Proceedings of the Academy of Natural Sciences of Philadelphia,* 20: 174–176.

Leidy, J. 1869. The extinct mammalian fauna of Dakota and Nebraska, including an account of some allied forms from other localities, together with a synopsis of the mammalian remains of North America. *Journal of the Academy of Natural Sciences of Philadelphia,* 2: 1–472.

Lemon, R. R. H., and C. S. Churcher. 1961. Pleistocene geology and paleontology of the Talara region, Norhtwest Peru. *American Journal of Science,* 259: 410–429.

Leyhausen, P. 1979. *Cat Behavior: The Predatory and Social Behavior of Domestic and Wild Cats.* Garland STPM Press, New York.

Lund, P. W. 1842. Blik paa Brasiliens dyreverden för sidste Jordomvaeltning. Fjerde Afhanlinger: Fortsaettelse af Pattedyrene. *Det Kongelige Danske videnskabernes Selskabs Naturvidenskabelige og Mathematiske Afhandlinger,* 9: 137–208.

MacCall, S., V. Naples, and L. Martin. 2003. Assessing behavior in extinct Animals: was *Smilodon* social? *Brain, Behavior and Evolution,* 61: 159–164.

Marean, C. W. 1989. Sabertooth cats and their relevance for early hominid diet and evolution. *Journal of Human Evolution,* 18: 559–582.

Marshall, L. G., and R. L. Cifelli. 1990. Analysis of changing diversity patterns in Cenozoic land mammal age faunas in South America. *Paleovertebrata,* 19: 169–210.

Martin, L. D. 1980. Functional morphology and the evolution of cats. *Transactions of the Nebraska Academy of Sciences,* 8: 141–154.

Martin, L. D., J. P. Babiarz, V. L. Naples, and J. Hearst. 2000. Three ways to be a saber-toothed cat. *Naturwissenschaften,* 87: 41–44.

Martin, L. D., V. L. Naples, and J. P. Babiarz. 2011. Revision of the New World Homotheriini. In V. L. Naples, L. D. Martin, and J. P. Babiarz (eds.), *The Other Saber-Tooths: Scimitar-Tooth Cats of the Western Hemisphere,* 185–194. Johns Hopkins University Press, Baltimore.

Martin, L. D., and C. B. Schultz. 1975. Scimitar-toothed cats, *Machairodus* and *Nimravides,* from the Pliocene of Kansas and Nebraska. *Bulletin of the University of Nebraska State Museum,* 10: 55–63.

Martin, L. D., C. B. Schultz, and M. R. Schultz. 1988. Saber-toothed cats from the Plio-Pleistocene of Nebraska. *Transactions of the Nebraska Academy of Sciences,* 16: 153–163.

Martínez-Navarro, B., and P. Palmqvist. 1995. Presence of the African machairodont *Megantereon whitei* (Broom, 1937) (Felidae, Carnivora, Mammalia) in the lower Pleistocene site of Venta Micena (Orce, Granada, Spain), with some considerations on the origin, evolution and dispersal of the genus. *Journal of Archaeological Science,* 22: 569–582.

Matthew, W. D. 1910. The phylogeny of the Felidae. *Bulletin of the American Museum of Natural History,* 28: 289–316.

Mazak, V. 1970. On a supposed prehistoric representation of the Pleistocene scimitar cat, *Homotherium* Fabrini, 1890 (Mammalia; Machairodontidae). *Zeitschrift für Säugetierkunde,* 35: 359–362.

McHenry, C. R., S. Wroe, P. D. Clausen, K. Moreno, and E. Cunningham. 2007. Supermodeled sabercat, predatory behaviour in *Smilodon fatalis* revealed by high-resolution 3D computer simulation. *Proceedings of the National Academy of Sciences of the United States of America,* 104: 16010–16015.

McKenna, M. C., and S. K. Bell. 1997. *Classification of Mammals above the*

Species Level. Columbia University Press, New York.

Meachen-Samuels, J. 2012. Morphological convergence of the prey-killing arsenal of sabertooth predators. *Paleobiology,* 38: 1–14.

Meachen-Samuels, J., and W. J. Binder. 2010. Age determination and sexual dimorphism in *Panthera atrox* and *Smilodon fatalis* (Felidae) from Rancho La Brea. *Journal of Zoology,* 280: 271–279.

Meade, G. E. 1961. The saber toothed cat, *Dinobastis serus. Bulletin of the Texas Memorial Museum,* 3: 23–60.

Méndez-Alzola, R. 1941. El *Smilodon bonaerensis* (Muñiz): estudio osteológico y osteométrico del gran tigre fósil de la pampa comparado con otros félidos actuales y fósiles. *Anales del Museo Nacional de Historia Natural "Bernardino Rivadavia," Ciencias Zoológicas,* 40: 135–252.

Merriam, J. C., and C. Stock. 1932. *The Felidae of Rancho La Brea.* Carnegie Institute of Washington, Washington.

Miller, G. J. 1969. A new hypothesis to explain the method of food ingestion used by *Smilodon californicus* Bovard. *Tebiwa,* 12: 9–19.

Miller, G. J. 1980. Some new evidence in support of the stabbing hypothesis for *Smilodon californicus* Bovard. *Carnivore,* 3: 8–26.

Milner, R. 2012. *Charles R. Knight, the Artist Who Saw through Time.* Abrams, New York.

Modesto, S. P., and N. Rybczynski. 2000. The amniote faunas of the Russian Permian: implications for late Permian terrestrial vertebrate biogeography. In M. J. Benton, M. A. Shishkin, D. M. Unwin, and E. N. Kurochkin (eds.), *The Age of Dinosaurs in Russia and Mongolia,* 17–34. Cambridge University Press, Cambridge.

Mones, A., and A. Rinderknecht. 2004. The first South American Homotheriini (Mammalia: Carnivora: Felidae). *Comunicaciones Paleontológicas del Museo Nacional de Historia Natural y Antropología,* 35: 201–212.

Morales, J. 1984. Venta del Moro: su macrofauna de mamíferos y bioestratigrafía continental del Mioceno Mediterráneo. PhD dissertation, Universidad Complutense de Madrid, Madrid.

Morales, J., M. Pozo, P. G. Silva, M. S. Domingo, R. López-Antoñanzas, M. A. Álvarez Sierra, M. Antón, C. Martín Escorza, V. Quiralte, M. J. Salesa, I. M. Sánchez, B. Azanza, J. P. Calvo, P. Carrasco, I. García-Paredes, F. Knoll, M. Hernández Fernández, L. van den Hoek Ostende, L. Merino, A. J. van der Meulen, P. Montoya, S. Peigné, P. Peláez-Campomanes, A. Sánchez-Marco, A. Turner, J. Abella, G. M. Alcalde, M. Andrés, D. DeMiguel, J. L. Cantalapiedra, S. B. A. García Yelo, A. R. Gómez Cano, P. López Guerrero, A. Oliver Pérez, and G. Siliceo. 2008. El sistema de yacimientos de mamíferos miocenos del Cerro de los Batallones, Cuenca de Madrid: estado actual y perspectivas. *Palaeontologica Nova,* 8: 71–114.

Morales, J., M. J. Salesa, M. Pickford, and D. Soria. 2001. A new tribe, new genus and two new species of Barbourofelinae (Felidae, Carnivora, Mammalia) from the Early Miocene of East Africa and Spain. *Transactions of the Royal Society of Edinburgh: Earth Sciences,* 92: 97–102.

Morlo, M., S. Peigné, and D. Nagel. 2004. A new species of *Prosansanosmilus:* implications for the systematic relationships of the family Barbourofelidae new rank (Carnivora, Mammalia). *Zoological Journal of the Linnean Society,* 140: 43–61.

Morse, D. H. 1974. Niche breadth as a function of social dominance. *American Naturalist,* 108: 808–813.

Mosser, A., and C. Packer. 2009. Group territoriality and the benefits of sociality in the African lion, *Panthera leo. Animal Behaviour,* 78: 359–370.

Muchlinski, M. 2008. The relationship between the infraorbital foramen, infraorbital nerve, and maxillary mechanoreception: implications for interpreting the paleoecology of fossil mammals based on infraorbital foramen size. *Anatomical Record,* 291: 1221–1226.

Murphey, P. C., K. E. B. Townsend, A. R. Friscia, and E. Evanoff. 2011. Paleontology and stratigraphy of middle Eocene rock units in the Bridger and Uinta Basins, Wyoming and Utah. In J. Lee and J. P. Evans (eds.), *Geologic Field Trips to the Basin and Range, Rocky Mountains, Snake River Plain, and Terranes of the U.S. Cordillera,* 125–166. Geological Society of America, Boulder, Colorado.

Naples, V. L. 2011. The musculature of *Xenosmilus,* and the reconstruction of its appearance. In V. L. Naples, L. D. Martin, and J. P. Babiarz (eds.), *The Other Saber-Tooths: Scimitar-Tooth Cats of the Western Hemisphere,* 99–122. Johns Hopkins University Press, Baltimore.

Naples, V. L., and L. D. Martin. 2000. Evolution of Hystricomorphy in the Nimravidae (Carnivora; Barbourofelinae): evidence for complex character convergence with rodents. *Historical Biology,* 14: 169–188.

Newman, C., C. D. Buesching, and J. O. Wolff. 2005. The function of facial masks in "midguild" carnivores. *Oikos,* 108: 623–633.

O'Brien, S. J., D. E. Wildt, M. Bush, T. M. Caro, C. Fitzgibbon, I. Aggundey, and R. E. Leakey. 1987. East African cheetahs: evidence for two population bottlenecks? *Proceedings of the National Academy of Sciences of the United States of America,* 84: 508–511.

Ochev, V.G. 2004. Materials to the tetrapod history at the Paleozoic-Mesozoic boundary. In A. Sun and Y. Wang (eds.), *Sixth Symposium on Mesozoic Terrestrial Ecosystems and Biota, Short Papers, 1995,* 43–46. China Ocean, Beijing.

Orlov, J. A. 1936. Tertiare Raubtiere des westlichen Sibiriens. I. Machairodontinae. *Akademia Nauk SSSR. Trudy Paleozoologiceskogo Instituta,* 5: 111–154.

Owen, R. 1846. *A History of British Mammals and Birds.* J. Van Voorst, London.

Packer, C. 1986. The ecology of sociability in felids. D. I. Rubenstein and R. V. Wrangham (eds.), *Ecological Aspects of social Evolution: Birds and Mammals,* 429–451. Princeton University Press, Princeton.

Packer, C., D. Scheel, and A. Pusey. 1990. Why lions form groups: food is not enough. *American Naturalist,* 136: 1–19.

Palmqvist P., J. A. Pérez-Claros, C. M. Janis, B. Figueirido, V. Torregrosa, and D. R. Gröcke. 2008. Biogeochemical and ecomorphological inferences on prey selection and resource partitioning among mammalian carnivores in an early Pleistocene community. *Palaios,* 23: 724–737.

Pascual, R., E. Ortíz-Jaureguizar, and J. L. Prado. 1996. Land mammals: paradigm for cenozoic South America geobiotic evolution. *Müchner Geowissenschaftliche Abhandlungen,* A30: 265–319.

Paula Couto, C. 1955. O "tigre-dentes-de-sabre" do Brasil. *Boletim do Conselho Nacional de Pesquisas,* 1: 1–30.

Peigné, S. 2003. Systematic review of European Nimravinae (Mammalia,

Carnivora, Nimravidae) and the phylogenetic relationships of Palaeogene Nimravidae. *Zoologica Scripta,* 32: 199–229.

Peigné, S., and S. Sen (eds.). In press. Mammifères de Sansan. *Mémoires du Muséum national d'Histoire naturelle.*

Petter, G., and F. C. Howell. 1987. *Machairodus africanus* Arambourg, 1970 (Carnivora, Mammalia) du Villafranchien d'Ain Brimba, Tunisie. *Bulletin du Muséum national d'Histoire naturelle de Paris,* ser. C, 9: 97–119.

Petter, G., and F. C. Howell. 1988. Nouveau Félidé Machairodonte (Mammalia, Carnivora) de la faune pliocène de l'Afar (Ethiopie): *Homotherium hadarensis* n.sp. *Comptes Rendus de l'Académie des Sciences,* 306: 731–738.

Pilgrim, G. E. 1913. The correlation of the Siwaliks with mammal horizons of Europe. *Records of the Geological Survey of India,* 43: 264–326.

Prothero, D. R. 1994. *The Eocene-Oligocene Transition: Paradise Lost.* Columbia University Press, New York.

Rabinowitz, A. R., and B. G. Nottingham. 1986. Ecology and behaviour of the jaguar (*Panthera onca*) in Belize, Central America. *Journal of Zoology,* 210: 149–159.

Radinsky, L. 1969. Outlines of canid and felid brain evolution. *Annals of the New York Academy of Sciences,* 167: 277–288.

Rawn-Schatzinger, V. 1992. The scimitar cat *Homotherium serum,* Cope. *Illinois State Museum Reports of Investigations,* 47: 1–80.

Reumer, J. F. W., L. Rook, K. Van der Borg, K. Post, D. Moll, and J. De Vos. 2003. Late Pleistocene survival of the saber-toothed cat *Homotherium* in northwestern Europe. *Journal of Vertebrate Paleontology,* 23: 260–262.

Riabinin, A. 1929. Faune de mammifères de Taraklia. Carnivore vera, Rodentia, Subungulata. *Travaux du Musée de Géologie de l'Académie des Sciences d'URSS,* 5: 75–134.

Riggs, E. C. 1896. *Hoplophoneus occidentalis. Kansas University Quarterly,* 5: 37–52.

Riggs, E. C. 1934. A new marsupial saber-tooth from the Pliocene of Argentina and its relationships to other South American predacious marsupials. *Transactions of the American Philosophical Society,* 24: 1–32.

Rincón, A., F. Prevosti, and G. Parra. 2011. New saber-toothed cat records (Felidae: Machairodontinae) for the Pleistocene of Venezuela, and the great American biotic interchange. *Journal of Vertebrate Paleontology,* 31: 468–478.

Ringeade, M., and P. Michel. 1994. Une nouvelle sous-espèce de Nimravidae (*Eusmilus bidentatus ringeadei*) de l'Oligocène inférieur du Lot-et-Garonne (Soumailles-France): étude préliminaire. *Comptes Rendus de l'Académie de Sciences, Paris,* ser. 2, 318: 691–696.

Riviere, H. L., and H. T. Wheeler. 2005. Cementum on *Smilodon* sabers. *Anatomical Record,* 285A: 634–642.

Robles, J. M., D. M. Alba, J. Fortuny, S. de Esteban-Trivigno, C. Rotgers, J. Balaguer, R. Carmona, J. Galindo, S. Almécija, J. V. Bertó, and S. Moyà-Solà. In press. New craniodental remains of the barbourofelid *Albanosmilus jourdani* (Filhol, 1883) from the Miocene of the Vallès-Penedès (NE Iberian Peninsula) and the phylogeny of the Barbourofelini. *Journal of Systematic Palaeontology.*

Roussiakis, S. J. 2002. Musteloids and feloids (Mammalia, Carnivora) from the late Miocene locality of Pikermi (Attica, Greece). *Geobios,* 35: 699–719.

Roussiakis, S. J., G. E. Theodorou, and G. Iliopoulos. 2006. An almost complete skeleton of Metailurus parvulus (Carnivora, Felidae) from the late Miocene of Kerassia (northern Euboea, Greece). *Geobios,* 39: 563–584.

Rudwick, M. 1992. *Scenes from Deep Time.* University of Chicago Press, Chicago.

Salesa, M. J. 2002. Estudio anatómico, biomecánico, paleoecológico y filogenético de *Paramachairodus ogygia* (Kaup, 1832) Pilgrim, 1913 (Felidae, Machairodontinae) del yacimiento vallesiense (Mioceno superior) de Batallones-1 (Torrejón de Velasco, Madrid). PhD dissertation, Universidad Complutense de Madrid, Madrid.

Salesa, M. J., M. Antón, A. Turner, L. Alcalá, P. Montoya, and J. Morales. 2010a. Systematic revision of the late Miocene sabre-toothed felid *Paramachaerodus* in Spain. *Palaeontology,* 53: 1369–1391.

Salesa, M. J., M. Antón, A. Turner, and J. Morales. 2005. Aspects of the functional morphology in the cranial and cervical skeleton of the sabre-toothed cat *Paramachairodus ogygia*

(Kaup, 1832) (Felidae, Machairodontinae) from the late Miocene of Spain: implications for the origins of the machairodont killing bite. *Zoological Journal of the Linnean Society,* 144: 363–377.

Salesa, M. J., M. Antón, A. Turner, and J. Morales. 2006. Inferred behaviour and ecology of the primitive sabre-toothed cat *Paramachairodus ogygia* (Felidae, Machairodontinae) from the late Miocene of Spain. *Journal of Zoology,* 268: 243–254.

Salesa, M. J., M. Antón, A. Turner, and J. Morales. 2010b. Functional anatomy of the forelimb in the primitive felid *Promegantereon ogygia* (Machairodontinae, Smilodontini) from the late Miocene of Spain and the origins of the sabre-toothed felid model. *Journal of Anatomy,* 216: 381–396.

Salesa, M. J., M. Antón, J. Morales, and S. Peigné. 2012. Systematics and phylogeny of the small felines (Carnivora, Felidae) from the late Miocene of Europe: a new species of Felinae from the Vallesian of Batallones (MN10, Madrid, Spain). *Journal of Systematic Palaeontology,* 10: 87–102.

Sardella, R., and L. Werdelin. 2007. *Amphimachairodus* (Felidae, Mammalia) from Sahabi (latest Miocene–earliest Pliocene, Libya), with a review of African Miocene Machairodontinae. *Rivista Italiana di Paleontologia e Stratigrafia,* 113: 67–77.

Schaub, S. 1925. Über die Osteologie von *Machairodus cultridens* Cuvier. *Eclogae Geologica Helvetiae,* 19: 255–266.

Schmidt-Kittler, N. 1976. Raubtiere aus dem Juntertiär Kleinasiens. *Palaeontographica,* 155: 1–131.

Schmidt-Kittler, N. 1987. The Carnivora (Fissipedia) from the lower Miocene of East Africa. *Palaeontographica,* 197: 85–126.

Schultz, C. B., M. R. Schultz, and L. D. Martin. 1970. A new tribe of saber-toothed cats (Barbourofelini) from the Pliocene of North America. *Bulletin of the University of Nebraska State Museum,* 9: 1–31.

Scott, W. B., and G. L. Jepsen. 1936. The mammalian fauna of the White River Oligocene: part I–Insectivora and Carnivora. *Transactions of the American Philosophical Society,* n.s., 28: 1–153.

Scrivner, P. J., and D. J. Bottjer. 1986. Neogene avian and mammalian tracks from DeathValley National Monument, California: their context, classification and preservation. *Palaeogeography, Palaeoclimatology, Palaeoecology,* 57: 285–338.

Seidensticker, J. 1976. On the ecological separation between tigers and leopards. *Biotropica,* 8: 225–234.

Seymour, K. L. 1983. The Felinae (Mammalia: Felidae) from the late Pleistocene tar seeps at Talara, Peru, with a critical examination of the fossil and recent felines of North and South America. MS thesis, University of Toronto, Ontario.

Shaw, C. A. 1989. The collection of pathologic bones at the George C. Page Museum, Rancho La Brea, California: a retrospective view. Supplement to issue 3. *Journal of Vertebrate Paleontology,* 9: 153.

Sigogneau-Russell, D. 1970. *Révision systématique des gorgonopsiens sud-africaines.* Éditions du Centre national de la recherche scientifique, Paris.

Simpson, G. G. 1941. The function of sabre-like canines in carnivorous mammals. *American Museum Novitates,* 1130: 1–12.

Sinclair, W. J. 1921. A new *Hoplophoneus* from the Titanotherium Beds. *Proceedings of the American Philosophical Society,* 60: 96–98.

Sinclair, W. J., and G. L. Jepsen. 1927. The skull of *Eusmilus. Proceedings of the American Philosophical Society,* 66: 391–407.

Solounias, N., F. Rivals, and G. M. Semprebon. 2010. Dietary interpretation and paleoecology of herbivores from Pikermi and Samos (late Miocene of Greece). *Paleobiology,* 36: 113–136.

Sotnikova, M. 1992. A new species of *Machairodus* from the late Miocene Kalmakpai locality in eastern Kazakhstan (USSR). *Annales Zoologici Fennici,* 28: 361–369.

Spassov, N. 2002. The Turolian Megafauna of West Bulgaria and the character of the Late Miocene "Pikermian biome." *Bolletino della Società Paleontologica Italiana,* 41: 69–81.

Spencer, L., B. Van Valkenburgh, and J. M. Harris. 2003. Taphonomic analysis of large mammals recovered from the Pleistocene Rancho La Brea tar seeps. *Paleobiology,* 29: 561–575.

Sunquist, M. E., and F. C. Sunquist. 1989. Ecological constrains on predation by large felids. In J. L. Gittleman (ed.), *Carnivore Behavior, Ecology and Evolution,* 220–245. Chapman and Hall, London.

Turner, A. 1984. Dental sex dimorphism in European lions (*Panthera leo* L) of the Upper Pleistocene: palaeoecological and palaeoethological implications. *Annales Zoologici Fennici,* 21: 1–8.

Turner, A., and M. Antón. 1998. Climate and evolution: implications of some extinction patterns in African and European machairodontine cats of the Plio-Pleistocene. *Estudios Geológicos,* 54: 209–230.

Turner, A., and M. Antón. 2004. *Evolving Eden: An Illustrated Guide to the Evolution of the African Large Mammal Fauna.* Columbia University Press, New York.

Valli, A. M. F. 2004. Taphonomy of Saint-Vallier (Drome, France), the reference locality for the biozone MN17 (Upper Pliocene). *Lethaia,* 37: 337–350.

Van Valkenburgh, B. 2007. *Déjà vu:* the evolution of feeding morphologies in the Carnivora. *Integrative and Comparative Biology,* 47: 147–163.

Van Valkenburgh, B., and T. Sacco. 2002. Sexual dimorphism, social behavior, and intrasexual competition in large Pleistocene carnivorans. *Journal of Vertebrate Paleontology,* 22: 164–169.

Vekua, A. 1995. Die Wirbeltierfauna des Villafranchian von Dmanisi und ihre biostratigraphische Bedeutung. *Jahrbuch des Römisch-Germanischen Zentralmuseums, Mainz,* 42: 77–180.

Viret, J. 1954. Le loess à bancs durcis de Saint-Vallier (Drôme) et sa faune de mammifères villafranchiens. *Nouvelles Archives du Muséum d'Histoire Naturelle de Lyon,* 4: 1–200.

Vrba, E. S. 1981. The Kromdraai australopithecine site revisited in 1980: recent investigations and results. *Annals of the Transvaal Museum,* 33: 17–60.

Wagner, A. 1857. Neue Beiträge zur Kenntniss der fossilen Säugthier-Ueberreste von Pikermi. *Abhandlungen der Königlich Bayerischen Akademie der Wissenschaften,* 8: 111–158.

Ward, P. D., J. Botha, R. Buick, M. O. De Kock, D. H. Erwin, G. H. Garrison, J. L. Kirschvink, and R. Smith. 2005. Abrupt and gradual extinction among Late Permian land vertebrates in the Karoo Basin, South Africa. *Science,* 307: 709–714.

Webb, S. D., B. J. MacFadden, and J. A. Baskin. 1981. Geology and paleontology of the Love Bone Bed from the Late Miocene of Florida. *American Journal of Science,* 281: 513–544.

Werdelin, L. 2003. Mio-Pliocene Carnivora from Lothagam, Kenya. In M. G. Leakey and J. D. Harris (eds.), *Lothagam: Dawn of Humanity in Eastern Africa,* 261–328. Columbia University Press, New York.

Werdelin, L., and M. E. Lewis. 2000. Carnivora from the South Turkwell hominid site, northern Kenya. *Journal of Paleontology,* 74: 1173–1180.

Werdelin, L., and M. E. Lewis. 2001. A revision of the genus *Dinofelis* (Mammalia, Felidae). *Zoological Journal of the Linnean Society,* 132: 147–258.

Werdelin, L., and R. Sardella. 2006. The "*Homotherium*" from Langebaanweg and the origin of *Homotherium. Palaeontographica,* 227: 123–130.

Werdelin, L., N. Yamaguchi, W. E. Johnson, and S. J. O'Brien. 2010. Phylogeny and evolution of cats (Felidae). In D. W. MacDonald and A. J. Loveridge (eds.), *Biology and Conservation of Wild Felids,* 59–82. Oxford University Press, Oxford.

White, T. D., S. H. Ambrose, G. Suwa, D. F. Su, D. DeGusta, R. L. Bernor, J.-R. Boisserie, M. Brunet, E. Delson, S. Frost, N. García, I. X. Giaourtsakis, Y. Haile-Selassie, C. F. Howell, T. Lehmann, A. Likius, C. Pehlevan, H. Saegusa, G. Semprebon, M. Teaford, and E. Vrba. 2009. Macrovertebrate paleontology and the Pliocene habitat of *Ardipithecus ramidus. Science,* 326: 87–93.

Witmer, L. M. 1995. The extant phylogenetic bracket and the importance of reconstructing soft tissues in fossils. In J. J. Thomason (ed.), *Functional Morphology in Vertebrate Paleontology,* 19–33. Cambridge University Press, Cambridge.

Wroe, S., M. B. Lowry, and M. Antón. 2008. How to build a mammalian super-predator. *Zoology,* 111: 196–203.

Yamaguchi, N., A. C. Kitchener, E. Gilissen, and D. W. Macdonald. 2009. Brain size of the lion (*Panthera leo*) and the tiger (*P. tigris*): implications for intrageneric phylogeny, intraspecific differences and the effects of captivity. *Biological Journal of the Linnean Society,* 98: 85–93.

Zdansky, O. 1924. Jungtertiäre Carnivoren Chinas. *Palaeontologica Sinica,* 2: 1–149.

Index

MAURICIO ANTÓN has painted paleomurals for the Sabadell Museum in Spain, the Museo de Ciencias Naturales de Madrid, the Florida Museum of Natural History, and the American Museum of Natural History. He has co-authored and illustrated numerous books, including *Dogs: Their Fossil Relatives and Evolutionary History*; *The National Geographic Book of Prehistoric Mammals*; *Evolving Eden*; *Mammoths, Sabertooths, and Hominids*; and *The Big Cats and Their Fossil Relatives*.

This book was designed by Jamison Cockerham at Indiana University Press, set in type by Jamie McKee at MacKey Composition, and printed by Martin Book Management.

The fonts are Electra, designed by William A. Dwiggins in 1935, Frutiger, designed by Adrian Frutiger in 1975, and Futura, designed by Paul Renner in 1927. All were published by Adobe Systems Incorporated.